John Taylor was born in 1931 and educated at
grammar schools and at Christ College, Cambridge
from 1950-56. He was a brilliant mathematics scholar
and gained a Ph.D. He has held many Research
and Visiting Fellowships at Universities and
Colleges in Britain and America, as well as on the
Continent. He held Chairs of Physics at the
University of Southampton and at Rutgers
University, New Jersey, and is at present Professor
of Mathematics at King's College, London. John
Taylor was also trained as an actor, and he has
performed in plays and films as well as directing
stage productions in Oxford and Cambridge. He has
had articles published in numerous scientific
journals, and his books include *The Shape of Minds
to Come* (1971) and *The New Physics* (1972). He has
also written several science fiction plays, and is well
known to millions from his appearances on radio
and television in interviews, discussions and
documentary programmes.

BLACK HOLES

The End of the Universe?

JOHN TAYLOR

FONTANA/COLLINS

First published by Souvenir Press 1973
First published in Fontana with addition of postscript 1974

Copyright © John Taylor 1973
Postscript copyright © John Taylor 1974

Printed in Great Britain by
William Collins Sons & Co Ltd Glasgow

Contents

Acknowledgements

This book owes its inspiration to the work of many scientists over several decades, who have been trying so hard to unravel mysteries of the black hole and its relevance to the universe in which we live. Naturally enough the work is most indebted to Albert Einstein, the father of relativity. Then come the pioneers who tried to discover the implications of Einstein's monumental work—Schwarschild, Landau, Chandrasckhar, Oppenheimer, Volkov, Serber and many others. Over the last decade the black hole, and what it stands for, has been especially probed by John Wheeler, ably assisted by the penetrating work of Penrose, Thorne, Misner, Hawking, Carter, Gibbons, Hoyle, Zeldovitch, Novikov, Ruffina, Sciama, Israel and many others too numerous to mention. To all of these I acknowledge my debt, and especially to my colleagues Chris Isham, David Robinson, Paul Davies, Dennis Sciama and Roger Penrose for very helpful discussions on gravity, black holes and cosmology. Above all, I would like to thank my good friend Ates Orga for his continued advice during the writing of this book.

Preface

This book is an attempt to explain one of the most important developments in science of this century – the advent of the black hole. Inside this object (formed by the collapse of a heavy star to such a condensed state that nothing, not even light, can escape from its surface) the fundamental laws governing our Universe appear to be destroyed, along with our usual concepts of space and time.

The black hole not only puts the scientific world in turmoil but also challenges many of man's basic ideas about his surroundings and his place in them. The implications of the black hole for man are as important, if not more so, than for science. They are specially relevant to man's attempt to grapple with the unknown and push his powers of reasoning to the ultimate when he tries to answer the basic questions of life and death, of animate and inanimate matter.

The black hole brings us face to face with the mysteries of the world. That is why the first three chapters of the book are concerned with trying to assess exactly what these mysteries are and what attitude to take towards them. The following three chapters introduce the black hole, discuss its observational evidence and describe how to use it as a source of energy. The surprises and horrors of the black hole interior are then considered. The death of intrepid explorers in its interior is a very minor aspect of the possibilities, which include time travel and visiting other universes.

At this stage it is possible to consider how new ideas brought by the black hole and of highly condensed matter can give new understanding of the beginning and end of the Universe, as well as answer a string of puzzles about historical records of the past.

The book closes with a discussion of how the black hole

must cause a radical change in our understanding of many concepts so long cherished by man – immortality, reincarnation, dialectic, space, time, mind, the Universe itself and certainty in that Universe. In total, the black hole requires a complete rethinking of our attitude to life. It leads to the liberating conclusion: in the beginning there was no beginning; in the future there will be no future. Man is at one with the black-hole Universe.

1 The Worship of the Unknowable

Mankind has, ever since he began to think, worshipped that which he cannot understand. As millennia have passed he has understood an ever-increasing amount about the world around him. He has even hoped, in his most optimistic moments, to comprehend it all. Yet man is now in the position of facing the ultimate unknowable, which can never be penetrated as long as he remains in his present physical form. That ultimate unknowable is the black hole. However hard he may struggle he will never be able to get out of this most fearsome object of the heavens once inside it. Nor can he ever find out what is happening in its interior if he stays outside, fearing to make the one-way trip.

Man's worship of the mysterious in the past has always bred decay and destruction. In the light of this, how are we to face up to the black hole? Are we to worship it also from afar, or should we *this* time try to comprehend it seriously?

The constantly augmenting knowledge of the world has only been achieved by centuries of dedicated work by men of science. Their efforts have been so successful that man's way of life has been changed out of all recognition by the technological application of their discoveries. From the cradle to the grave the results of science are used to provide an artificial environment giving safety and comfort from the buffetings of nature. It is as if man departs from his mother's womb to enter straight into another one created by the scientists.

Even before birth new drugs are used to help the foetus survive. Once born and for the whole of his life medical discoveries allow him to drug himself, have bad parts cut out of his body, or good ones transplanted into it.

His final demise can be long delayed. He may even spend

his dying years as a vegetable, his body being almost completely run by automatic machinery keeping his blood circulating, his lungs still breathing, his vital organs functioning correctly and his bowels satisfactorily empty. The equipment to run his life is powered by forms of energy only recently discovered. Electricity, especially, has allowed him to turn night into day and let him communicate across vast distances with others of his own sort.

When he has grown safely to adulthood he can wake up in the morning in his heated or air-conditioned house, use the latest techniques to prepare food for himself, drive off in his heated or air-conditioned car, and spend the day in a glass and plastic office using electronic methods of communication to spread the consumption of such advanced technology ever farther over the earth's surface or advance it to ever more highly developed forms. If he wishes he may travel at faster than the speed of sound to any point on the globe, or even exceptionally be one of the select few who have voyaged to the moon. And to cap it all he may, if he really so desires, stay at home and change into a she!

To obtain an ever faster increase in the rate of scientific advance man is forcibly exposed in his earlier, more vulnerable years, to the wonders of this scientifically contrived world to which he is heir. Such universal education is regarded as his birthright and woe betide the unwary who try to remove it, or deny it to certain sections of the population. And indeed the headlong acceleration of scientific discovery would be stopped dead in its tracks if there were not an assured supply of the educated coming forward to offer their lives and join the ranks of the high priests and practitioners of science.

All this, and so much more, has been done in the name of science and technology. Yet in spite of these many-coloured wonders of the world the first item of news that the average person turns to in the newspaper is not of the latest scientific discovery or natural disaster, or even of the most recent political shenanigans. It is the astrology column. The average man or woman wants to know what fate the stars have in store for them. Even though there is absolutely no scientific basis for it, astrology is gaining great support in the Western world; in the East it has always had strong foundations. 'Palaces of Astrology' are opening in the main cities of Western countries. Throughout the earth people are increasingly looking to the stars for supernatural influences.

This growing support for astrology is only one of many pointers to a general disenchantment with science and all that it stands for. There is greater interest everywhere in anti-scientific activities, as evinced by the general hippy and drop-out culture of the last decade, with its bizarre music, resort to drugs and great interest in the literature of fantasy. A world inhabited by Hobbits, with their dislike of machines more complicated than a forge-bellows, a watermill or a hand-loom, and by Dwarves, Elves, Wizards and terrible creatures of the dark has more appeal to many of the younger generation than one in which the atoms wend their certain way to unleash ever-greater power for man over his surroundings. If Middle Earth was ever found on this earth by a lucky explorer it would be deluged by the vast number of humans wishing to live there. There is also the fascination of the world of towers and crags of Gormenghast, full of its ritual and symbolism. Many would dearly love to be its seventy-eighth lord.

Fantasy may even take so great a control that strange objects and visitors from outer space are seen and, in extreme cases, communicated with. Unidentified flying objects are still of great popular interest, and it is felt by some that there is a conspiracy afoot among the hard-headed establishment to smother reports on such contacts and bring the world back to its senses. An alternative fantasy is the world beyond the senses, of the 'psi' faculty – telepathy, telekinesis, precognition and other powers of the mind. This activity merges into witchcraft, which is still practised by a surprisingly large number of educated people in the world.

Even greater numbers turn to orthodox religions of various sorts, all of them appealing to supernatural authority of some form or other to find meaning in life. As science has advanced it is to be expected that religion would have retreated. But apparently, while that may have occurred to a certain extent

13

in the early stages of the Scientific revolution, it no longer does at anything like the same rate. Religion is even gaining popularity in some countries, surprisingly in the most technologically advanced of all – America.

We are naturally led to ask what produces this increasing revolt against science and the life of reason. There would, in fact, appear to be a general cause for this. We live in a world which is certainly incomprehensible. We ask how and why we are here, what was the beginning of it all, what will its end be, what is time or space, and many other impossible questions. If we keep our eyes closed and our heads down and refuse to worry about these whys and wherefores of our existence we can usually muddle through. But once we look up and notice how strange our world really is then we very soon find ourselves drowning in a sea of ideas, conjectures, wild surmises and beliefs. Each of these pulls us towards a different island of hope and assurance like so much flotsam and jetsam. Many of us finally acquiesce and establish ourselves on the best island of faith we can find. Some of us try to struggle on, but often glance with envy at those who have swallowed the carrot of a particular set of beliefs.

But even those securely ensconced in their faith's replies to the ultimate questions of existence are still aware of problems. The answers given to the faithful are vague and full of ambiguity. And they may not always be as satisfying as they were at first glance. Growing up and maturing, being faced by new difficulties to be resolved and decisions to be made, the old beliefs may be seen to be inadequate. They may even conflict directly with recent scientific discoveries, and great psychological stress can be caused to the faithful by this discrepancy. Yet the need for an answer to the problem of existence is a strong force in the modern world. We have to feel we understand our environment, so that we are better prepared to face any threats it may present to us.

The security a faith provides seems, on the surface, to be of the same sort as that provided by science. To the faithful, events happen the way they do because God, alien beings, the stars, or other remote powers cause them to. As with science there is the same apparently causal chain of explanation. But these remote powers to which the fanatics appeal possess two much more attractive features not contained in the scientific world of cause and effect.

Many of us are impatient. Life is short, and we wish to have as much satisfaction as we can obtain as rapidly as possible. In particular we want a reason here and now for why the world is the way it is. We have to make crucial decisions in our lives which depend upon this world picture. We cannot wait for a lengthy chain of causes to be discovered by the patient analysis of science, especially because the longer the chain the farther removed it becomes from the human sphere of relevance. We might appreciate that matter is made up of atoms, but when the atom is found to have a nucleus inside it and that nucleus has elementary particles inside it which themselves have quarks inside them, well, we tend to lose track and interest. This is especially so when the quarks are to be expected to have internal constituents, which will themselves have components, and so on in a never-ending succession.

So scientific explanation does not really seem an explanation at all because it never ends. There are always further steps of the explanation waiting to be uncovered. This is just the opposite to the famous American President Truman who is said to have had a notice on his desk stating 'The buck stops here'. And so it does for the faithful, stopping at their God or whatever stands for it. A man's act of faith can carry the buck for him for the rest of his life provided it is strong enough. Science seems incapable of doing so.

The second feature that makes faith score over science is that it can be all-embracing and explain every aspect of the world, not just a severely limited part of it. The scientific method only applies to reproducible phenomena, and the effort to achieve this repeatability often drives phenomena away completely or at best distorts them severely. This is especially so as far as human experiences are concerned, and imagination, creativity, feelings and indeed any mental experiences would seem to be well beyond the scientific pale. An example of this is the act of love observed under the cold gaze of laboratory scientists; in the words of the poet 'we lost the love some time ago, now we've only the act to grind'. This limitation leaves the major part of human experience apparently out of bounds to scientific scrutiny. So faith steps gracefully in and explains all.

A rapid, all-embracing framework with which to explain the hostile Universe makes faith and fantasy far more appealing than science in this short life of a mere human. Even scientists may turn to clutch at religious straws as their lives thunder on toward the final whirlpool. The certainty they searched for during their too-short lives could only be obtained in the last few years, days or minutes of it by rejecting the scientific code they lived by up till then.

There is another feature of science which gives ground to the other side, and may well cause this breakdown of the scientific spirit. As understanding of the world increases ignorance also expands, for there are fresh problems which are exposed to view by the new advances. It is as if our known world lies on the surface of a balloon, with the unknown just on the outside of it. As the balloon expands so the known world increases, but so does the area outside it with which it comes into contact; our ignorance is simultaneously expanded. This increasing ignorance of our surroundings can finally sap the confidence of an ageing scientist ever to understand the world by scientific analysis. This difficulty is also seized upon by the anti-scientific to show how little we understand about the world by the help of science. How much easier it is to appeal to supernatural forces in some form or other!

Naturally enough the powers possessed by the Gods of the faithful are awesome. They evoke feelings of reverence in the minds of their devotees. To prevent these forces being used against the believers themselves, the Gods must be propitiated. So services of worship of the unknowable Gods are developed and permeate the whole of life. By verbal expression of man's insignificance in the mind of God, of his weakness in the face of the Almighty, and of other comparisons, the faithful places the unknowable in a position from which it no longer threatens his life. He has manipulated it so that it is powerless to harm him.

This process of defusing the unknowable was evidently

essential in the pre-scientific age when man felt so much more at the mercy of the natural elements. But it is surprising to find it still practised so extensively when he has the weapons of the technological revolution in his hands. These weapons evidently have failed to neutralize the threat of the unknown.

However hard we try the inexplicable will not disappear. It will always loom large in our lives, and can only be propitiated by belief in the supernatural unless we are prepared to look it full in the face and square up to what precisely is strange, what cannot be explained. In doing this we have to reject our preconceived notions and throw away the security of a particular position of faith. We must bring to our attack on the unknown all the knowledge of the past and of the present, and all of the hopes of the future. Only then can we begin to gain any understanding at all of the impossible surrounding us. Not that we may get very far, but even a step or two into the gloom of the unknown could be worth a million miles travelled with the blinkers of faith over our eyes.

If we do not attempt the perilous journey into the infinite, fraught as it is with unknown dangers of which nightmares are made, we could well end up with a civilization devoted to the worship of the unknowable. And that would be of the utmost danger to each and every one of us, for it could spell the end of our culture as we know it, and even of the human race itself.

We can appreciate this better if we look at past civilizations which flourished on earth for a limited period and then decayed away. For example, the Sumerian, Egyptian and Aztec cultures all persisted for many centuries. They reached great heights in the arts and practical sciences, yet in none of them did the understanding of the world around them advance very far. For over two millenia the Sumerian civilization, for example, was the cradle of culture in the Middle East, and developed advanced legal and agricultural systems. Yet the complete dependence of the Sumerians on the Gods of the temple and of nature caused blinkers to be put over men's eyes. Their whole lives were guided by the signs and omens of the Gods. The security and tranquillity afforded by the advanced society of Sumeria was not sufficient to allow its inhabitants to develop sufficient control over the forces of nature to withstand the onslaughts of neighbouring, less developed, societies. While the Sumerians worried if the omens of the Gods were propitious the uncaring barbarians raped and pillaged their cities and destroyed their civilization.

A similar situation arose for the Aztecs. They had developed a correct calendar, a system of counting which included zero and high architectural and artistic skill. Theirs was an urbanized society, with the city of Tenochtillan containing over three hundred thousand people at the time of the Spanish conquest in the sixteenth century. Yet their lives were also dominated by ritual and worship of their Gods. This undoubtedly had a highly debilitating effect on the culture. Despite a strongly militaristic society the Aztecs were finally dominated by a band of about five hundred armoured Spaniards led by Herman Cortes. The Aztecs were to a certain extent destroyed by their own religion, since they initially accepted Cortes with open arms as their white God, Quetzalcoatl, making a prophesied return.

These are only few of the many examples from man's past which show that worship of the unknown can reduce his capability to survive below the critical value. Devotion to ritual and dogma lead to loss of flexibility and the power to adapt to a hostile environment. Only once in man's history has he effectively thrown off the shackles of the fear of the infinite and attempted to face himself and the world and get closer to the true nature of things.

It was a poet, the Italian Petrarch, who can be credited with this important advance. Through his poetry and writings, especially about the ancient Greek thinkers, he sowed the seeds which flowered into the Renaissance in Europe. Before then the life of medieval man had been ruled and shaped by the social and intellectual traditions of his family, his guild, his feudal class and his church. Renaissance man rejected such restrictions on his acts and thoughts. He was free to do what he wanted and to think as he pleased. He could also look at his surroundings in a new light and more fully savour the beauty of nature as it related directly to himself. This breakdown of the reliance on dogma in Europe ushered in the scientific revolution, and allowed Western man within the last three centuries to gain the power over nature which has led us to where we are today, in the technological society.

We have now completed a full circle and returned to where we began. But our perspective is enriched, so that we can look ahead with the wisdom of hindsight. The view of the future, at least of the next few decades, is not a pretty one. To the threat of the unknowable to the scientific society must be added its increased vulnerability as it becomes ever more complex and the stresses under which it is placed increase in severity.

Much has been written about the dangers from the three P's: pollution, population, and power (where the latter is of the energetic and not the political sort, though that, also, is fraught with risks). Some of these writings are no doubt exaggerated for sheer effect, but there is undoubtedly a crisis facing the modern world due to the three P's. Reduction of pollution to manageable levels must necessitate increased manufacturing costs, all needing a great deal of effort and creating many stresses. The worst pollutant is man. He is increasing his numbers at so fast a rate, doubling them in the next thirty years, that ever-decreasing amounts of essential raw materials and too slowly increasing quantities of food will produce more and more people at or below the subsistence level. Simultaneously, the world's power requirements are far out-stripping her capacity; the hungriest giant of all, America, has already had to curtail some essential services such as schools and airline services due to dwindling supplies of oil.

As each year passes the strains on the fragile fabric of society produced by these contradictory forces get steadily greater. At the same time society becomes more vulnerable with ever greater dependence on complicated methods of distribution of the essentials for life. Natural forces can still cause great disruption, since only one or two key installations need be destroyed for chaos to ensue. Modern societies know this only too well, being, for example, at the mercy of determined workers in key industries who decide to strike for higher wages. A case in point occurred recently in the United

Kingdom when disruption of electricity supplies almost brought the country to a standstill; strike power increases in proportion to the complexity of society and its concomitant fragility.

No mention has been made of the spectre of the nuclear holocaust, still flitting hauntingly through the corridors of power. This only makes the problem worse; when the crash comes it will very likely bring radioactive fall-out with it to eliminate mankind.

When we take account of these various threats facing us the future looks bleak indeed. It is impossible to foretell the path ahead with any certainty, as kings and politicians have found so often to their cost. But undoubtedly the next five decades will be the most testing time which mankind has ever experienced. The fate of *homo sapiens* hangs in the balance. How can racial suicide be avoided?

The rituals of the past have never saved a civilization from destruction. Nor can we expect a return to one or other of those rituals to save mankind now. There might be a future ahead for humanity if all worshipped in the same form, and believed in the same unknowable powers controlling the universe, whatever they were. Let us even discount the enormous variety that would undoubtedly exist in such a society, ultimately leading to its instability and destruction. But it does not appear even faintly possible to achieve such uniformity of belief in the short space of time available to us. Five decades is at most three generations; mankind cannot change that fast.

It is too late to return to the ordered days of pre-Renaissance Europe, or to the omen-directed lives of the Sumerians. Too much power over nature has been forged by man ever to give it back. Too many ideas have been thought ever to return to the Age of Innocence, whenever that was – if at all. The only hope is to turn to science to get mankind out of the present chaos, especially because science can be blamed for a lot of it. The population explosion has been ignited by the blessings of modern medical science, and the industrial destruction of the environment by new knowledge in other branches of science and technology.

These essentially scientific disciplines, controlled by wise and thoughtful leaders, are man's only hope of solving the problems of the three P's over the next few decades. But they can only do so if there is whole-hearted support for such a procedure. The swing from science towards faith in and worship of the unknowable must be halted. The tide must be turned to show that science is adequate to face the mysteries of existence. It must be demonstrated that a complete scientific view of the world can be obtained which is psychologically as powerful and satisfying as that given by the supernatural powers – the Gods, alien beings or all-powerful stars – that are currently increasing their followers.

We have to see that science can rend the veil from the face of the temple of the unknown. Even if we only find another veil underneath, at least we must continue to probe. For as each covering is removed we may discover that the unknowable takes a form before our eyes that is no longer threatening. We may see a picture of a world at harmony with us and with all things, controlled by principles of the utmost simplicity. This is an alternative method of defusing the mysterious to that of simply labelling it 'Infinite' and kneeling before it. It has the advantage over the latter of not only giving us security but also greater understanding of our environment and ourselves. Hopefully that will enable us to pass through the coming testing period with the least danger of destruction.

We must, of course, be prepared to find that the face of the unknowable is even stranger than we might have expected. Our voyage of discovery may reveal a world totally different from the naïve one we see around us every day. In one sense that will make life harder, since the threat of the infinite will be even greater. But in another fashion it will smooth the way ahead, for the forces, once glibly labelled 'supernatural', that we will ultimately face will be completely different from those which most of us see around us in the world today.

We already have evidence in the heavens of objects of most bizarre character. We are now realizing that the forces producing them can create the supreme bizarre object. This is the ultimate unknowable. It is the black hole.

2 The Inexplicable

As we turn to face the awe-inspiring mysteries of our existence we realize that one of the most difficult of these is time, that ever-flowing stream. As Omar Khayyám so aptly wrote:

> 'The moving finger writes; and, having writ,
> Moves on: nor all your Piety nor Wit
> Shall lure it back to cancel half a line
> Nor all your Tears wash out a Word of it'.

We are born but to die, in this never-ceasing stream. No one has ever been able to dam it up, nor even slow it down. We all will have as life-span the Biblical three-score years and ten, if we are lucky, and then we depart this earthly experience.

We ask insistently: what is time? We fantasize over it, we imagine that we can build machines allowing us to travel through time, and the still high popularity of H. G. Wells' science fiction story *The Time Machine* shows how many of us would like to sit at the controls of such an instrument. But none has been made, and present-day science does not appear to have any blue-print for such a one on its drawing boards.

Yet there is evidence in the past of far longer lives than our presently allotted span. Chapter 5 of Genesis records ages of the earlier generations on earth which are ten times or more than present ones. Adam is said to have lived for nine hundred and thirty years, his son Seth nine hundred and twelve, his grandson Enos nine hundred and five, while Methusaleh, nine generations after Adam, lived to the ripe old age of nine hundred and sixty-nine years. These are all remarkably close together and so could be explained by a common biological feature which they transmitted till it later died out. But may they have known how to control time 'from outside', so to say, this still later being lost for some reason? Of course we don't know and we may never know for certain. That is no reason for not attempting to conjecture how such control could have been achieved. Was time better understood by the ancients than ourselves? Did they have a knowledge of it denied to us with our greater sophistication? Possibly we just cannot see the wood for the trees.

The same questions arise when we consider many other aspects in which time is crucially involved. One closely related to the ability to control the flow of time is that of precognition – being able to see into the future. It has been experienced at all levels, from the dream which comes true to the prophecy of a whole chain of events. A famous example is that of J. W. Dunne, who dreamed one night in the autumn of 1913 that he was looking from a high railway embankment at a scene which he recognized as being situated a little to the north of the Firth of Forth Bridge. Below him was open grassland on which groups of people were walking about. The scene came and went several times, but on the last time he noticed that a train going north had fallen over the embankment. He saw several carriages lying near the bottom of the slope, down which large blocks of stone were rolling. He then warned his friends not to travel north by train to Scotland. On April 14, 1914, the 'Flying Scotsman' jumped the parapet near Burntisland Station, about fifteen miles north of the Forth Bridge and fell onto the golf links twenty feet below.

Many people have experienced similar dreams which come true later; I have myself. It is very difficult to explain this phenomenon with our present comprehension of time as being outside our control. This precognition seems to show that we may have a slight power of changing time; in particular in sleep. We now know that sleep is a very active process, and especially that our brains may be working harder then than while we are awake, so such a possibility cannot be ruled out at once.

Precognition, as well as time travel into the past, bring into question another feature of the world around us that we take for granted unless we think about it carefully. This is causality, that every event is preceded by some prior event which we call the cause, the subsequent event its effect. The cause of a person's death can be one of a number of events – an accident, a malfunction of a critical organ for no apparent reason, the murderous action of someone else, and many other possibilities. But why should every event have a cause, and moreover why should cause always precede effect? Can we never be in the situation in which a window breaks before a ball is thrown at it, or a man falls dead before his enemy shoots at him? Because we do not normally experience such a curious phenomenon here on earth does not mean that it can never happen on this globe, nor that there are not regions of the Universe in which causality is not normally true.

In fact we might have to find out if there are even more bizarre places if we want to be able to turn time's flow backwards in a time machine. For if we voyage to the past then we must cause some disturbance in the region we visit; if we don't we cannot see or sense in any way whatever what is going on around us. Any change caused by the time traveller could be enough to alter later events in a very unexpected fashion.

But then he could get himself into some very awkward situations, such as disturbing his past so badly that it develops into a future which doesn't contain him at all. This looks an impossible situation to be in, though one whose possibility we have to face if unrestricted time travel is to be considered. We will evidently have to try and sort it out properly before we can safely turn our time machine on!

There is another way in which causality and time come into conflict with our common sense ordering of the world, and that is as to the nature of the very early and late stages of the world. Here the mystery of our existence becomes clearly exposed, and our minds begin to reel under the heavy onslaught on our reasoning ability. If there is a beginning to the world there must have been a first event. But what happened before that? And if there is an end to it all what will take place afterwards? On the other hand, the Universe may have no beginning or no end. But such an eternal existence is well-nigh impossible to comprehend; the human mind is only finite.

Most religions attempt to avoid the occurrence of an ever-existing world by introducing a time of creation, before which there was nothing, after which there existed at least the seeds of our present Universe. However, the explanation is only a side-step which still leaves us open to the same difficulties. For we may ask what caused this act of creation. The answer is usually that it was an act of God, or of the Gods if there are a plurality in the religion. In Genesis, 'In the beginning God created the heaven and the earth'. But who created God? If the answer is nobody, then we require an eternally existing deity, otherwise we can ask who created God's creator, his creator, and so on.

God, the infinite and unknowable creator of all, can only have had no other creator if God had been able to create both the Universe and God himself. So we can only reach a satisfactory state of affairs for the beginning of the world if we have a very powerful God indeed, able to create himself, as well as us, out of the previous absence of any quantity or quality whatsoever. That is a feat indeed worthy of the infinite, but one which seems unlikely. For how does he start to create himself? He has absolutely nothing to begin with, to get a toe-hold, so to speak, for he has no toe before he has any existence at all. At the same time we can ask even if the idea of time itself has any sense when there is absolutely nothing around that can be used to mark it off with. Before God's self-creation there are no events to follow time's ever-flowing stream. Indeed there is very likely no time if there are no events. So how did God know when to begin his remarkable act? And how could he even know when to know to begin, even if time made sense, if he didn't yet exist, so be unable to know anything?

We have, then, effectively two equally mysterious possibilities about the world's beginning: either it has always been here, or it created itself in some fantastic 'bootstrap' process, before which it doesn't make any sense to talk about time.

There are similar possibilities about the world's end: either it is snuffed out, and time and existence have no further meaning, or it continues forever, at least in some form or other. There is the same difficulty about being annihilated into nothing as being created from nothing. However, a world going on for ever appears to be easier to swallow than one which has always existed. The belief in immortality is widespread and still strongly held. For it to be possible it is essential that the Universe, or at least that part of it where the immortals reside, must itself always exist. This is borne out by the deathless Gods in the majority of the earth's religions.

Not all of man's Gods live for ever. It would be more natural to expect that if man has created God in his own likeness he would make him mortal. This has indeed been the case in a number of cultures. Thus the Greenlanders believed that a wind could kill their most powerful god and that he would certainly die if he touched a dog. North American Indians believed the world was made by the Great Spirit, but that he had died long since, not having been able to survive for so long. Heitsi-ebib, a god or divine hero of the Hottentots, died several times and came to life again. His graves are generally to be met within narrow defiles between mountains.

The Greek gods were not exempt from this over-all decay. The grave of Zeus was shown to visitors in Crete as late as the beginning of our era, while the body of Dionysus was buried at Delphi beside the golden statue of Apollo, and his tomb bore the inscription 'Here lies Dionysus dead, son of Semele'. Neither were the gods of Egypt any different from man; they too grew old and died. At a later stage the idea of immortality arose and the gods were allowed to share this with man. It is usually suggested that this invention came about through the discovery of embalming, the continued existence of the physical body without deterioration giving hope that it might be brought back to life at a later stage. This certainly seems to be the case for the later Egyptian gods, such as Osiris, whose mummy can be seen at Mendes, or Anhouri at Thinis.

It would indeed be surprising if the idea of immortality came only through a technological development. But yet where else could it have arisen from to have taken hold of men's minds so strongly that it was dominant in the later stages of the earth's cultural development, and especially in the world's main religions of Christianity, Judaism, Buddhism, Hinduism and Mohammedism? The puzzle presented here is not impossible, but one to be solved, if at all, by the use of all the relevant evidence from the past, by our understanding of the present, and by our utmost utilization of our imagination. This will hopefully allow us to conjure up a vision as to the set of events which could put deep into men's hearts such a strong belief in immortality and spread it across the world.

There are further puzzles embedded in the mesh of religious beliefs engirdling the earth. The gods created man in their own image, so man universally claims. As we read in Genesis i, 26: 'And God said, let us make man in our image, after our likeness.' The 'Teaching of Ani' which appears to date from towards the end of the Egyptian Empire says categorically that 'Man is the counterpart of God', and humans are described as 'replicas of God which issue from his limbs' in the 'Menims of Khety'. This attitude is very puzzling, since how can the all-powerful, infinite be duplicated and replicated many times? We know that we are very puny beings at large in a dangerous world. It is obviously healthy to have as patron the Lord and Master of it all. But why have to be made in his image?

The mystery deepens when further ancient records are read. Again from the Christian Bible in Genesis vi, 1-2, 'And it came to pass, when men began to multiply on the face of the earth, and daughters were born unto them, That the sons of God saw the daughters of men that they were fair; and they took them wives of all which they chose'. And again in Genesis vi, 4: 'There were giants in the earth in those days; and also after that, when the sons of God came in unto the daughters of men, and they bore children to them, the same became mighty men who were of old, men of renown.' This tradition of contact and even interbreeding between God's relatives or God himself and mere humans was strongly developed in the Christian religion to lead to its present form which is completely based on the son of God, Jesus Christ, born of the union of God and a human female. How could God have sons, unless He were in human or near-human form, or at least possessed of the ability to perform genetic experiments on mankind. The 'sons of God' are interpreted by scholars as other Gods in the ancient pantheon, this being the sense used, for example, in a Canaanite text of the eighth century B.C. But there is still the problem as to how these lesser Gods themselves achieved such intimate relations with women. Should we consider these 'sons of God' as very powerful near-human beings come possibly from outside the earth?

To the believer, the mysteries of how such events could occur and who these sons or son of God were are not to be discussed further, but worshipped for themselves. These problems may well have power to attract so many humans now precisely because of the extreme difficulty in penetrating the mysteries expressed by the Biblical writings. It is the infinite, the patently unknowable, which is at the heart of these doctrines, and it is that to which people now bow their heads on a Sunday in their place of worship, or why they lower their voices when passing the local vicarage. There is indeed what we might call an event horizon round the religious doctrine. Outside the horizon we can obtain no information from the inside; outside is the non-believer, who cannot appreciate the descriptions of the infinite contained in the doctrine of the faith inside the horizon. If he believes, he goes inside the horizon, where he experiences oneness with the unknown. While this may not be very illuminating it can be extremely satisfying. The believer can hear what the non-believer is saying and understand it, but cannot get the non-believer outside to understand what he is saying.

It is often very difficult to say that there is no belief in the supernatural in a civilization or a particular group of people. The early anthropologists and explorers were especially deceived about religious belief among primitive tribes. For example, the famous explorer Sir Samuel Baker said, in 1866, that the Northern Nilotes were without any belief in a supreme being, nor had they any form of worship. This has since been shown quite false; statements about people's religious beliefs must always be treated with caution.

We can no longer leave one of the most difficult questions of all: from where does the experience of God spring? All primitive cultures and most developed ones have expressed a relationship with one or more super-human beings. Only recently have completely atheistic societies come into being which do not profess to such beliefs. The Gods of the past and present form a bewildering array of shapes and sizes, powers and weaknesses. But yet they all have a special relation with human beings, and have often been in direct communication with them.

Naturally man must be able to influence his Gods and vice versa. If he cannot change their actions by suitable propitiation then they are not really his Gods helping him survive in an alien world. This may take quite an extreme form, such as human sacrifice, which has persisted in modified form even till very recently in the sacrificial slaying of a cock, ram or lamb in many important acts, such as the building of a house. The Aztecs even went so far as to slaughter nearly one hundred thousand captives on one religious occasion. If they did not have any victims handy they would even go and start a war for the purpose of capturing them; a bloodthirsty God indeed!

If we go to the earliest evidence of man's religion we find mainly the worship of the female, especially when she is pregnant. Archaeologists have dug up a whole galaxy of female figures from about 4500 B.C., usually of naked women standing or seated, and often pregnant. The earliest temple compounds in the whole world, dating from 4000-3500 B.C. in or near Mesopotamia appear to have been dedicated to the worship of the Mother, two of these even being of oval form designed to suggest the female genitalia. The first recorded God was thus not a God at all but a Goddess.

Within about a millenium the force of the female had been reduced, and male Gods began to take ascendance; these have been superior ever since. It is usually explained that this change came about through the development of more complex civilization based on the city and not the primitive village, along with development of skills which removed husbandry from its pre-eminent position. But there is another possibility which we should not reject out of hand, especially since it has recently been put forward by reputable scientists. The crucial idea is that mankind's development has been influenced, possibly strongly, by visitors from extra-terrestrial civilizations. Since it looks very unlikely that there can be intelligent life on our sun's other planets, such intervention can only have come from a planet circling a star at least several light years away. Can these space-beings have caused the change from the worship of the female to make the male at least equal if not superior? Indeed could these visitors be the sons of God we referred to earlier, mentioned in Genesis?

Besides these rather brief records in Genesis there are other passages in the Bible which are difficult to explain rationally without calling upon some intervention of beings from outer space, as has been done by the Soviet ethnologist, M. M. Agrest. He has suggested, for example, that the account of the destruction of Sodom and Gomorrah is very close to that expected from an observer living in ancient times with no understanding of nuclear explosions. The description in Genesis xix, 27-28 reads as follows: 'And Abraham got up early in the morning to the place where he stood before the Lord: And he looked toward Sodom and Gomorrah, and toward all the land of the plain, and beheld, and lo, the smoke of the country went up as the smoke of a furnace.' This description fits quite well that of an atomic explosion, as would the reason why Lot's wife should not turn back and look at the spectacle. She did and would have been blinded, though turning into a pillar of salt must be metaphorical.

There are even quite graphic accounts in the Bible describing the visitation of alien space-craft. Certain of these involved some type of craft which carried God or his angels and made a tremendous noise, emitting clouds of smoke. The prophet Ezekiel gives a very clear description of this: 'And I looked and, behold, a whirlwind came out of the north, a great cloud, and a fire infolding itself, and a brightness was about it, and out of the midst thereof as the colour of amber, out of the midst of the fire.' Ezekiel goes on to describe the details of the craft and that 'I heard also the noise of the wings of the living creatures that touched one another and the noise of the wheels over against them, and a noise of great rushing'. He then narrates how the craft spoke to him, and how he was even given what was apparently a stimulant to 'pep him up'.

There are numerous similar references. Some of these give vivid descriptions of space-craft, such as that which guided Moses and the Israelites out of the land of Egypt. 'And the Lord went before them by day in a pillar of cloud to lead them the way; and by night in a pillar of fire, to give them light; to go by day and night' (Exodus 13:21). A pillar of cloud by day and a pillar of fire by night is a good description of a rocket as seen by a primitive person who has never even seen an aeroplane. Moses was also warned of the physical dangers of the space-ship, in Exodus 19:11-12: 'And be ready against the third day: for the third day the Lord will come down in the sight of all the people upon about Mount Sinai. And thou shalt set bounds unto the people round about, saying, Take heed of yourselves, that ye go not up into the mount, or touch the border of it: whosoever toucheth the mount shall be surely put to death.'

There are also numerous accounts in other ancient records of space-craft visiting the earth and having contact with mankind. The Tibetan books Tantyua and Kantyua also mention prehistoric machines, calling them 'pearls in the sky'; the Samerangana Sutrodhara has whole chapters describing airships whose tails spout fire and quicksilver. The sacred Indian epic, the Mahabharata, dating back to about 3000 B.C., has various descriptions of flying machines, which could cover vast distances and could travel forwards, upwards and downwards, with enviable manoeuvrability.

These visitations from the heavens have apparently persisted over the millenia, if more recent records are reliable. Nor do they seem to have ceased in present time, and were particularly plentiful in the late 1940's. An especially tragic case was that of Captain Thomas Mantell who was killed in 1948 when his P-51 fighter plane disintegrated after he had given chase to a huge gleaming metal object, which he said '. . . looks metallic – and it's tremendous in size'. He found himself below it and was very close to it when his plane disintegrated with the falling wreckage scattered over thousands of feet.

Since then there have been many more sightings of strange objects in the sky, though no definite identification of these phenomena as powered craft has been achieved. This leads us to ask why there has been no definite contact with the beings controlling such space-craft if that is what the objects are. If they have been visiting us over such a long period of time why have they not been to the centres of our civilization and opened up a two-way conversation?

This may well have happened in historic times, as has been recently suggested by the American astronomer and exobiologist Professor Carl Sagan. He has remarked that legends of the origin of Sumerian civilization categorically state that non-human intelligent beings contacted the Sumerians during various periods of time and imparted their wisdom to help the civilization progress. This is especially interesting since it is considered that Sumer was very likely the first civilization, in our present understanding of the word, on our planet, dating back to the fourth millenium B.C. or even earlier. So our own level of progress may owe itself to such contacts.

The meeting appears to have taken place on the shores of the Persian Gulf, possibly near the site of the ancient Sumerian city of Eridu. There are three separate accounts of it which agree with each other, and they can all be traced back to a priest, Berosus, who lived in the city of Babylon at the time of Alexander the Great. He had access to cuneiform and pictograph records going back to several thousand years before his own time. The three accounts say that a being called Oannes with the body like that of a fish, though under the fish's head another head like a man's, and feet like a man's joined to the fish's tail, came out of the Persian Gulf and conversed with men. It gave them an 'insight into letters and sciences and every kind of art', according to one of the accounts. Similar beings appeared at later stages, and these arrivals mentioned exactly the things told by Oannes. It is also interesting to note that the kings of the time were supposed to have lived for very long periods of time until the coming of the flood; at that time a deity is also supposed to have warned the then king of its imminence and to safeguard the knowledge then extant.

How much should we believe these ancient records? Are they just figments of the imagination, valiant attempts to explain the unknowable, but seen from our superior vantage point evidently incorrect? Interestingly enough it has not proved easy to explain the rather abrupt transition from chaos to civilization at the beginning of recorded time in Sumeria. The Danish-American Sumerologist Thorbild Jacobsen of Harvard has written 'Overnight, as it were, Mesopotamian civilization crystallizes. The fundamental pattern, the controlling framework within which Mesopotamia is to live its life, formulate its deepest questions, evaluate itself and evaluate the Universe, for ages to come, flashes into being, complete in all its main features'. There has been recent evidence of a more gradual change, but the difficulty is still present in a lessened form. And along with it the cause of the quite rapid change from Goddesses to Gods. All caused by Beings from Outer Space? We don't know, but it may be so.

There is some curious evidence in support of the idea of extraterrestrial visitations in the illustration on small cylinders, called cylinder seals, which when rolled on clay would leave an impression. These pictures carry clear representations of celestial objects, as a circle or sphere, surrounded by smaller similar objects. The number of these smaller objects is variable; they would seem to be planets encircling a star, but why do their number vary? The number of planets round our sun was not known in those days, and the very idea of planets circling sun and stars did not originate till Copernicus many thousands of years later. So how could there be pictures of various planetary systems so long ago? Or were they various stars in communication with each other, as they might be nearer the Galactic Centre where stars are more closely packed than for us out in a spiral arm? Sumerian mythology supports this, the Gods being celestial in origin, each usually associated with a different star, and governed by a group of prominent deities called 'the Seven Gods Who Determine Destinies'. As Carl Sagan has noted, such a picture is not altogether different from what we might expect if a network of confederated civilizations interlaced the Galaxy.

Of course if mankind has been helped by alien beings there are as many new questions that need to be answered as are solved. How did they get here? How did they achieve the long life they would seem to need if the accounts in the Bible or the earlier Sumerian records are anything to go by? And how did they help man lengthen his own life span, at least in selected cases? Why have there been no open contacts with the more developed human civilizations of recent times, even if we accept the earlier meetings as true? Are we to expect future visits, and if so, when? Will these visitations be friendly, especially now that we have powerful atomic weapons as a possible threat to alien beings? All of these questions and so many more spring to mind; we will try to answer some of them later on when we have assembled more evidence, especially of a more concrete sort.

There are still many inexplicable things which lie in wait to trip us up whenever we feel ourselves a little superior. The list is still long, though the ones we have ticked off – time, precognition, causality, the beginning and end of the world, the reason for the belief in God, and visits from outer space, form an important part of the unknowable. We have not touched on the problem of mind – is it different from matter or are they two aspects of an underlying reality? This is a question which has been especially discussed in the last few centuries, the final answer not yet in sight. Contact with other intelligent beings might help clarify it, or they may even have resolved it. They may possess advanced telepathic powers, or other abilities demonstrating mind as supreme. Nor have we considered the problem of space, and especially does it go on for ever or, instead, stop somewhere? But, if so, then what is on the other side of the edge? The problem is the modern version of that about the earth, is it flat or round? We know the answer to that one, or at least most of us do, though the 'Flat Earthers' still make their presence felt occasionally, as the following story shows.

A friend of mine, working at a Services research laboratory, received a letter from a flat-earther who said that he had secret information that the Russians had planned to set off an atomic bomb at their edge of the earth which would cause it to tip up and, being unprepared, we would fall off. What were the Services going to do about it? My friend, being kind, decided to humour the writer, and replied that they knew all about this and also had an atomic bomb, placed at the opposite edge, which they would explode at the same instant as the Russians. A week later my friend was called in to his commanding officer's room and given a severe dressing-down for revealing secrets of research in the lab. Apparently the flat-earther had written to the commanding officer praising my friend's perspicacity and recommending him for promotion.

It is not helpful to believe blindly in flat earths, nor in many of the things which might make life much simpler and more secure. We have to look at all the available evidence and see what it tells us. We must do this even about the infinite, that which we really feel, in our heart of hearts, we will never be able to understand. It is through careful analysis of what we see around us, essentially by the scientific method, that we can hope for at least a little advance into the unknown. So let us turn to see how far science has drawn back the curtains over the mysteries of life. We can then start to look at the pictures which we are beginning to see exposed to our view. In particular we will try to discover how far science can take us towards the infinite.

3 The Optimists

Most of us are pessimists in life. We usually buy far more life insurance than we really need. Some of us even take out special coverage when we fly by plane 'just in case', though the statistics indicate we are more at risk within a few minutes of getting in a car, and drugs of all sorts are used by a large section of the population with an alarming number of people dying from an overdose. There are many in society who are depressed and require expert psychological care to alleviate their unhappiness. Such reactions are natural considering that we are bombarded on all sides by the news media with stories of the general unpleasantness of life. Pain and violent death seem to be all around us. No wonder that we search for a way out, trying to escape the horrifying realities of life. We flee into fantasy, mysticism, witchcraft, sex and drugs, leaving reality behind to care for itself.

Yet there have always been some amongst us prepared to look the real world in the face. They have even attempted to understand its inner workings. Beside man's innate curiosity there is one very practical reason why this has proved worthwhile to them and given a reasonable return for the pain of seeing the infinite as it really is. Ever since the beginning of recorded time the mysteries of life have provided power for those with daring and imagination to interpret them according to their own fashion. A little knowledge of heavenly or earthly natural processes gave their possessors even greater power over their credulous fellows. This understanding was naturally jealously guarded to preserve the control its holders had gained. It was also surrounded by ritual to preserve its mystery and add to its power. It was in this way that early knowledge of nature was developed as part of flourishing religions.

There is a very clear example of this in ancient Egypt, about five to six thousand years ago. Then the astronomer priests could predict accurately when the River Nile was going to overflow, since that occurred on the day that Sirius, the Dog star, rose above the horizon just before dawn. This knowledge was available to everyone who could observe it, but since it would have needed records preserved over many years, the priests' secrecy over their observations made such independent observations unlikely.

The manner in which this was used was beautifully expressed recently by the distinguished biologist-historian Professor Homer Smith, who wrote 'The man who first discovered the Sirius-Nile cycle was doubtless a tongue-in-the-cheek shaman who took advantage of his discovery to impress everyone with his esoteric power. He had in his possession knowledge never before available to any human being, for he could tell the people when to prepare the ground and plant the seed, when the flood would begin, and when it would subside. Because he knew so much it appeared that he knew everything, and he quickly convinced his fellow-Egyptians that it was his rites and ministrations that, by appealing to the gods, sustained these rhythms. He alone could mediate between men and gods because he alone knew the histories and wishes of the supernal beings'. This power became so great that ultimately a priest became king in Egypt.

This use of natural knowledge to gain power over one's fellow-men has persisted but has changed radically in its manner of operation. There is now a complete separation between the awe-inspiring mysterious, and the known and understood phenomena; the first is still worshipped under the name of religion, the second under the name of science. In religion one has to be initiated with the faith and encouraged to make the leap of belief to accept the mysterious for what it is, as always mysterious. The scientist is trained in completely the opposite tradition, always to question how, always to penetrate the unknowable and make it clear, and always to spread this knowledge as well as he can to his colleagues. The scientist of today can indeed be regarded as the extreme optimist in the face of the infinite. He does not accept that a particular problem can never be understood. To him all is possible, all will appear clear if only he works hard enough and well enough.

The scientific optimists are now hard at work trying to penetrate the unknown. But it isn't always easy for them. In the past the guardians of mystery have protected their citadel with all the means at their disposal. This has meant silencing the curious by various forms of punishment, ranging from verbal attack to the final one of death. The more powerful the guardians the more extreme the punishment.

In ancient Greece the preservers of the traditional, deified concept of the world fought very hard to combat the atheistic notions of the early Greek thinkers. Ananagoras, who believed that the moon was lit up by the sun, as well as being inhabited like the earth, was arrested in Athens in 434 B.C. on a charge of denying the godhead of the sun and the moon. He was lucky to escape with his life, owing it to the eloquent pleading of his friend Pericles, but he was still banished from Athens. Democritus, the father of the idea of the atom as the indivisible particle of matter, probably escaped condemnation for such heresy by keeping well away from Athens, then the centre of the worship of the Gods. The worst case of Athenian justice was that of the great philosopher Socrates, who was executed in 399 B.C. because he was guilty of 'firstly, denying the gods recognized by the state and introducing new divinities, and secondly of corrupting the young'.

Even in those early days the thinkers were optimists trying to understand the universe. One of them, Ananagoras, deduced from a large meteorite which fell in Greece in 467 B.C. that the sun was a mass of molten iron. He also interpreted eclipses correctly, and had the idea that heavenly bodies were formed by condensation from a chaotic whirling mass. All this is very close to modern ideas, as was the explanation of the heavens as a sphere with either the earth or the sun at its centre.

Each one of us has had to go through the trauma in early childhood of discovering that we are not the centre of the Universe, that there are other people, especially our parents, whose wishes are as important, if not more so, than our own.

We can still regain a little of that lost sense of uniqueness by having our earthly home as the focus of existence, and man as the supreme form of living being. This attitude is particularly attractive to religions which regard man as a special creation of the Gods, put on earth for some essential reason. Early Christian geographers drew their maps so that Jerusalem was the centre of the earth and the Holy Sepulchre was shown at the centre of Jerusalem. The man-centred attitude was also claimed to prevent men from living on the other side of the earth, even if it was allowed to be spherical. For if they did they would not be able to see Christ, at his second coming, descending through the air. It was for holding a contrary opinion that the astronomer Cecco d'Ascobi was burnt by the Inquisition in Florence in 1327.

It was only by a long and painful struggle of the scientific optimists against such religious strictures that man was finally removed from having any relation to the central point of the Universe. It took hundreds of years, a brutal killing and many imprisonments, but the truth won out in the end. We now know that we are the puny inhabitants of an average planet moving round an average star some distance away from the centre of an average galaxy (or collection of stars), itself in an average cluster of galaxies. This picture has only finally been clarified within the last twenty years, but now it seems certain that our physical place of habitation in the Universe has absolutely nothing special about it at all.

Not only do we have no cause for pride about where we live but also about who we are. Human beings are physically insignificant in the extreme. The vast and lonely reaches of space in which we live can only inspire in us fear; the extreme forms of matter at the centres of stars or in the waste of intergalactic space can only induce a feeling of awe. We are as nothing in an unthinking, unfeeling Universe.

This lesson is hammered home when we consider the size of our planetary system in comparison with the distance from our sun, to our nearest star, to the centre of our galaxy, and to our nearest neighbouring galaxy. Pluto, the most distant planet, is about one billion miles away (a billion being a thousand million), while Proxima Centauri, the nearest star to our sun, is over six thousand times as far away again. Such truly astronomical distances are best expressed by the time taken for light to traverse them at the fantastic speed of one hundred and eighty-six thousand miles in one second. At such a rate it still takes light about four years to reach us from Proxima Centauri, while it takes over thirty thousand years to speed across the Milky Way to us from its turbulent centre. The nearest galaxy to our Milky Way is about two hundred thousand light years away, whilst the further objects yet observed are thought to be as far away as five billion light years or more. In miles, that distance is three followed by twenty-two noughts!

Nor is space the only thing to make us feel truly minute when we try to see ourselves in the right perspective in the Universe at large. The number of stars in our own galaxy of various types and ages is more than one hundred billion, while the total number of stars in the observable Universe is roughly one hundred billion billion, these being contained in over one billion galaxies like our own.

The lengths of time involved in the evolution of stars and galaxies are also truly astronomical. Whereas man's recorded history has only lasted for five thousand years and our species, *homo sapiens*, for about twenty times as long, the sun has apparently been in existence for five billion years and our galaxy was probably formed about ten billion years ago. There are indications that the whole Universe itself was created in some 'big bang' from a dense primordial blob some two billion years earlier.

Nor does the catalogue of immensity end here, since there are enormous variations of conditions under which matter exists in various parts of the Universe. For example in interstellar space there may be as little as one atom in a cubic centimetre, while normal matter in stars is often about as dense as water with about a million billion billion atoms in a cubic centimetre. But this is still rarefied compared to the nucleus of an atom, where nearly all of its mass is concentrated. Nuclei are composed of varying numbers of two equally massive particles, the proton and neutron, the first having a positive electrical charge, the other none. It is difficult to compress nuclei together because they always have a net positive electrical charge which pushes them apart from each other. If matter could be made only of neutrons then such repulsion would not occur. Under certain conditions, especially in very massive stars, it is thought that purely 'neutronic' matter can exist. The central part of very heavy stars is expected to be almost exclusively made of such matter; it is so dense that one matchbox of it would weigh billions of tons.

Temperatures also range from the very lowest to the very highest imaginable, being almost absolute zero in intergalactic space yet millions of degrees at the centres of visible stars where thermonuclear processes are taking place. Energy is another quantity of great variation in the Universe, from our species burning fossil fuels and beginning to tap the power of the nucleus, to our sun pouring out one hundred billion times more power than that generated by humanity. But that is chickenfeed compared to the energy emitted by a star when it goes through the highly explosive period known as a supernova. During this cataclysmic process the star may increase its brightness up to one hundred million times its usual value, and may even shine more strongly than the entire galaxy which contains it.

Man appears truly dwarfed by this catalogue of immensity. How can he preserve any optimism in the face of it, or resist going into a deep depression? And even if he could begin to understand the processes at the basis of the unfolding of the Universe going on before his eyes how can he ever obtain any effective control over them? On the face of it, we all might as well give up and join the pessimists worshipping the infinite surrounding us. At least we should accept the inevitable gracefully instead of continually bucking it.

Recent developments have allowed us to be a little more hopeful that we might be able to play somewhat more of a role than our puny physical size would indicate. The only factor we evidently have to our advantage is our intelligence for what it is worth. And that will be a very great deal indeed.

We are the only species on this earth which is intelligent enough to develop language and writing, allowing us to build upon our fathers' and forefathers' wisdom. It is this which has produced the scientific revolution of the last three centuries, so allowing us to become the superior animal on our own planet. It has also allowed us to look at ourselves more closely, and especially our own thought processes. We come here to another of those impossibles which the traditionalists try to keep swathed in mystery clothes. What is thought and mind? Is it God-given, never to be understood? Or will we be able to build machines which we can really say have mental processes, and even consider ourselves as one of them which might be improved on?

We might expect the fight over that one to be over now, since it was effectively settled in 1859 with the publication of Darwin's 'Origin of Species'. Darwin's picture of man's evolution from more primitive animals by 'survival of the fittest' caused an enormous reaction. The battle raged very violently in the succeeding few years, especially in the public debate in June 1860 between the biologist T. H. Huxley for Darwin and Bishop Wilberforce of Oxford for God. The Bishop ridiculed Darwin's thesis, ending by sarcastically enquiring of Huxley whether it was through his grandfather or his grandmother that he claimed descent from a monkey. Huxley presented a masterly review of the evidence for evolution and finished by saying that he would rather have an ape as an ancestor than an intellectual prostitute like the Bishop (at which a Lady Brewster achieved immortality by fainting and having to be carried out).

But decidedly the battle was not over. The side of God only withdrew a little to lick their wounds and regroup. While man's physical body is now almost universally accepted as having evolved from lower animals, his mind does not come in the same category. Even some of the earth's foremost research workers into this problem do not accept that the mind is only a very useful development of the brain and that mental processes are completely determined by physical reactions proceeding in that complex organ. But the only attitude that the optimist can take is precisely that the brain explains the mind. This is proving more reasonable as the fantastically complex control of all features of mental processes by the brain are slowly becoming elucidated.

The more extreme, antisocial acts of sex perverts and compulsive aggressives or gamblers are even now being cured by brain surgery or the use of suitable drugs; dreams and other parts of sleep, fear, past memories and a whole host of sensations can now be controlled by electric currents passed into suitable parts of the brain. This developing power over the mind and the resultant reduction of freewill is realized as posing enormous problems for society in the future, and already breaking into criminal law. But this is only the thin end of the wedge; there is no apparent obstacle which will prevent the physical control of brain over mind to halt its inroads. Ultimately we can expect that all of the mental world will be reduced to the physical.

To many such a position is not optimistic at all but the most pessimistic one possible. They bewail the loss of spontaneity and of artistic inspiration if man is ever truly reduced to a machine. He will become nothing more than a robot, so they say. This is utter nonsense: man will always preserve his beautiful powers of creativity, of constructing works of art which will give great emotional satisfaction to his fellowmen. Just because we know how to cure certain forms of mental disease does not mean that a particular patient has become a robot. In fact he has usually become a more effective human being after his treatment. And that is precisely what we can look forward to for humanity by means of the new understanding being gained by what we call the mental revolution of the last hundred years.

The hope of the optimists is that man will use his intelligence to improve himself considerably and so speed up his understanding and control of both himself and his surroundings. This will not happen at the very slow rate of natural evolution, with its changes only being noticeable over a million years or so, but over decades or even shorter periods. It has already resulted in the construction of artificial aids to our intellectual processes that have allowed us to send space-craft to the moon and the neighbouring planets of our sun. A very great deal has already been achieved; even more will be understood about the mind in the near future.

The progress made in understanding our physical surroundings in this century, and especially in the last few decades, is even more astounding. Man has now explained the behaviour of matter over distances so small as to be incomprehensible even twenty years ago. He has also begun to construct a model of how the whole Universe behaves, out to the present limits of observation of many billions of light years, which is based on the properties of the sub-microscopic atoms and their constituents. The world of the very small has proved of crucial importance to that of the very large.

This unified view of the Universe is making the impossible look much more possible. We now understand the basic properties of matter in terms of four basic forces and several basic constituents. These forces are the attractive one of gravity between any pair of massive bodies, the electromagnetic force between electrically charged bodies, the nuclear one holding the neutrons and protons together inside the nucleus, and finally the force responsible for the radioactive decay of heavy atoms such as uranium. This last force we can call the 'weak force' because it is so much weaker than the electromagnetic and nuclear ones.

The constituents of matter between which these four forces act have been made ever smaller as their constituents have subsequently been found. Matter is made up of molecules, the smallest unit having the same properties as the matter they constitute. Different types of matter – solid, liquid or gas – have the molecules arranged in different ways, being packed close together with very little motion in a solid, having more movement in a liquid and almost complete freedom in a gas. These molecules are themselves made up of atoms, which were thought by the Greeks to be the ultimate indivisible units of matter. We know that they are the smallest components which preserve their chemical properties, but they can be broken down further. Atoms consist of electrons, which are light negatively charged particles, orbiting round a central, massive nucleus, itself composed of neutrons and protons. The various chemical properties of the atom are determined by the number of its orbiting electrons while the mass is determined by the number of protons and neutrons in the nucleus.

There is no need that the chain: matter – molecule – atom – electron plus nucleus – electron, proton and neutron – should stop at this point. We can search for the constituents of the electron, proton and neutron. This has been actively pursued over the last two decades and though no definite answer has arisen yet there are hints from the excited companions to the proton and neutron which have been created by slamming these latter particles into each other at very high energy. It has been suggested recently that the constituents of neutrons and protons are further particles called 'quarks' but they have not yet been found. They may well turn up when the new generation of particle accelerators is in full operation within the next year, giving even greater powers to probe inside the electron, proton and neutron.

But then why should we stop at the quark if and when it is found? Why not search for the constituents of quarks, and then their constituents and so on *ad infinitum*? If this were to be the situation then after many millenia of following the trail of constituents of constituents of . . . we would undoubtedly become disillusioned and accept that the impossible can never be achieved. Scientists would then turn mystics and worship the unknowable, not trying to penetrate it further.

These is a possibility that the path down into the atom will not continue for ever. The reason for this is related to the four fundamental forces of nature – the gravitational, electromagnetic, nuclear and weak. How do we explain these; how do they arise? Modern physics allows us to understand each of them in terms of a particular particle either discovered over the last few decades or on the verge of discovery. Each force between two particles can be described as arising by the exchange between them of that special particle corresponding to the force. Thus the gravitational, electromagnetic, nuclear and weak forces arise from the exchange of particles called the graviton, the photon, the meson and the intermediate boson.

The relevance of this to the search for the ultimate constituents of matter is that whatever they are they will have to produce precisely the force fields we know and no others. It appears very difficult to introduce constituents for the elementary particles of nature – the electron, proton, neutron, meson, photon and so on – without changing their mutual interactions too radically. An alternative approach has thus been to cause the elementary particles to produce each other by exchanging each other; for this to work the forces between them would have to be correct. In this way they will be then their own constituents. This is like getting human beings to create each other by a child, once born, creating its own father and mother. This would be possible if an efficient time machine could be constructed so he could go back to engender his parents being his own grandpa, so to say. The approach has had some success, and evidently is the way to break the infinite sequence of constituents of constituents of . . .

This self-consistent situation has even been investigated for the Universe in the large, though only in a very tentative way as yet, because there are many complications when the inter-mingling of large and small is being considered. But certainly the world of atomic and nuclear physics has entered crucially into astrophysical research because of the discovery that energy generation in a star is by nuclear fusion, most essenti-ally when four atoms of hydrogen fuse to make one of helium and produce a suitably large amount of energy. But there is no reason why the bootstrap process cannot be extended to the Universe as a whole. If that were possible then the impossible would surely have been achieved. There could only be one Universe, the one we live in, engendering itself and all that is in it.

But we have to be careful that we have not been too optim-istic. It would seem that a self-consistent Universe would give us something for nothing. We have to start with some pro-perties or laws into which we fit this self-created world. We need to assume some properties about time and space, about energy and its various manifestations, about the law of cause and effect, and so on. But where do these assumptions come from? Can they engender themselves also, or do we have to take them for granted? If we are forced to the latter situation then we are really in no better situation than when we started on our quest for the ultimate. We would have met it, face to face, and realized that we can penetrate no further.

It is this possibility which we have to be prepared for in the future. If self-consistent explanation of the constituents of matter failed, and there was a continued sequence of constituents of constituents of . . . of matter we could follow this trail as long as our strength lasted. It would either continue indefinitely or ultimately cease producing new constituents; we would be left for ever with the same ones in the latter case. In either case there would always be mystery, either that which we could never know finally if there were always new constituents to be found of the current ultimate constituents of the Universe, or met with directly if there were not. On the other hand, unless we can bootstrap our fundamental concepts of space, time, etc., the self-consistent approach will still produce mystery, that contained in the basic inexplicable concepts.

The possibility of such ideas engendering themselves seems impossible. How can we even conceive of time being created self-consistently? We are back to the same problem raised earlier – how can God create himself? It just does not seem possible for him to do this. Instead we will have to accept that the world will always contain mystery. The impossible will always be there.

But we need not leave it at that. Though the optimists may be misguided in their belief in a completely explicable world they may still remain partial optimists. They can worship at the shrine of the mysterious only most effectively when they know as well as they are able what that mysterious is. It may well consist of the various undefined concepts like space, time and so on, together with some fundamental theory relating these ideas. The optimists will have gone as far as they can and will then recognize their limitations.

We come back to the question raised earlier. Why not be pessimistic in the face of this ultimately existing mystery? The best answer to give to that is that we must understand our surroundings to gain power over them. If we do not we may be snuffed out at some time in the future. The sun may suddenly explode, or as some indications give, may be cooling off quite rapidly. We may have a star in our neighbourhood undergo a supernova explosion and destroy life on earth by highly damaging radiation. We must be prepared for such eventualities, and work towards providing a suitable future for mankind. This will only come through the continued strivings of the optimists. They must discover all of the accessible secrets of the Universe, especially by looking at the most extreme situations. For it is under such conditions that our present comprehension is put to the most stringent test, and so advances. We must look around us on earth and in the heavens for the unexpected, the phenomena for which our present understanding proves inadequate.

And that is where the black hole plays a very strong role. It is in there that the laws of physics are in the melting pot, where time and space ends or begins. We must dive into it and see what it contains.

4 The Black Hole Appears

None of us wants to die, but death is with us whether we like it or not; sadly we have not yet discovered how to avoid it. All living things begin to die as soon as they are born; none can escape. Mortality may even be a crucial feature of being alive. Yet we still hope and search to find immortality, just as the alchemists looked so valiantly but vainly for it in the past. It is not usual to talk about the life and death of non-living things, so it would not seem to help our search for the elixir of life to consider them seriously. Yet some of these 'dead' things behave so similarly to live ones – they can be seen to change, to age, and even to decay to more primitive components – that we are naturally led to say they have been born, have grown old and eventually die. If we look carefully at some of these changes we might possibly be able to detect the non-living immortals, and by further investigation even obtain an understanding of how to slow time down, and ultimately to stop it altogether. So death may be avoided by analysing the 'never alive'.

In the past man persistently ascribed immortality to the stars – the immutable orbs which he thought controlled his destiny. Of course modern 'scientific' man now knows how naïve such an idea is. This is especially so since we discovered that some stars can suddenly flare up in a supernova explosion to become millions of times brighter, and then rapidly burn themselves out, never to be seen again. We also now realize that all stars must be changing since they are constantly giving out light from their surfaces and so losing energy. This energy has to come from somewhere, and also be in a limited supply; when all the star's fuel has been burnt up we expect the star to die out.

On the face of it, then, the stars are not immortal. But

ancient man may have been closer to the truth than he knew, since in the death of a star may lie the seeds of our own immortality. There may also be other far less pleasant surprises in store for us, and we might even be brought face to face with the truly infinite. For it is in the stars that the greatest energies and masses reside in the Universe; it is there, in the fiery interior of our sun or of the other stars sprinkled through the heavens that the fires of ultimate creation and destruction are raging. We must attempt to warm our hands at them, hopefully avoiding being reduced to a charred cinder in the process. As we advance we will realize even more why the greatest caution is necessary in our endeavour; the ultimate danger can be lurking ready to annihilate us.

If a star can grow old and die, it can presumably be born. Stars are even now forming throughout our Milky Way, as well as in a myriad of other galaxies throughout the Universe. They are made out of the wisps and traces of hydrogen gas and minute dust particles of graphite or sand which float around between the present stars. In our own galaxy the gas and dust is concentrated in spiral arms winding out from the galactic centre; shock waves travelling round the galaxy spark off the condensation of the interstellar material into large clouds or globules of matter. These can be seen as dark patches hiding the stars in various parts of the sky or as glowing nebulae excited by energy from the stars near them.

Once gathered together the condensing material collapses further, the various dust and gas constituents being drawn to each other irresistibly by the mutual gravitational attraction between them. Why the force of gravity between bodies is always attractive has never yet been fully explained, and it may be that under extreme conditions it becomes repulsive and allows us to build the much conjectured anti-gravity machine. But in the early stages of star genesis the condensing clouds are too thin for gravity's attraction to fail, and the clouds thicken, even drawing in further material as they do so.

Millions of years must pass before a collapsing cloud can be recognized as anything remotely resembling a star. But as the condensation proceeds the cloud begins to heat up, its constituents jostling each other with increasing vehemence as they are squashed closer and closer together. The cloud begins to glow, first of all with radiation not visible to the human eye but in the infra-red or with radio wavelengths; astronomers have recently seen some of these cocoons cradling the infant stars by means of infra-red and radio telescopes.

At this stage there may be one or even many objects in the collapsing cloud which are themselves sufficiently hot to be called stars, since they are shining with their own light. As they condense further they shine ever more brightly, and ultimately can be seen twinkling through the murky envelope of gas and dust surrounding them. The source of their light is their own collapse; it is as if they have to try to commit suicide to be born.

Yet they have hardly started their life-cycle. Further millions of years are required before each star has collapsed sufficiently for it to be so hot and luminous that its light can be seen by us, countless billions of miles away through space. And it is only then that a star really comes of age. For now its further collapse is halted by energy being produced at its centre battling its way to the exterior, in the process exerting enough pressure on the in-falling material to cause it to halt. The star has now truly appeared, and can begin its sometimes lengthy career.

It is worthwhile remarking here that the murky atmosphere surrounding a newly-formed star will in all likelihood continue condensing locally. The resulting dense objects will be too small even to become stars, so turn into planets. This means that most stars will have their own planetary systems; the number of them able to support intelligent life is expected to be over one million in our galaxy alone. The better we understand every detail of how the planets are formed the sooner will we know where to search for intelligent life in nearby stars. And without doubt there is someone out there, possibly even searching at this very moment for life like ours.

One of the puzzles facing man ever since the ancient Greeks, and even before, is where the stars get their energy. Only in the last century has it been realized how enormous is the amount of energy that each star will squander on its surroundings. For example, our sun pours out 4 million tons of its own mass every second, about ten thousand times the rate at which water in the River Thames flows under Waterloo Bridge. Only in the last two decades has any understanding of this energy generation really been arrived at. By the fusion of the nuclei of lighter elements to form heavier ones energy is gained through the small loss of mass involved. The star's initially high proportion of hydrogen allows it to build up helium, then carbon, oxygen, nitrogen and successively heavier elements, gaining energy in the process. At each new stage fusion produces more energy till a reasonable proportion of the star has been turned into iron. By then nuclear fusion no longer liberates energy; formation of elements heavier than iron by fusion of lighter ones is even an energy consuming process.

When a star has used up all its available lighter elements in this way it no longer has an interior source of energy sufficient to prevent its further gravitational collapse. It has nearly reached the end of its life as a stable star and starts a brief period of very dangerous activity, both for itself and all around it. It reaches this climacteric the more rapidly the heavier it is; our sun may live for several billions of years, while a star fifty times heavier might last only a thousandth of that time. The heavy central material of the star can even soak up some of the available energy, shoring the star up against its own collapse. So there may be a fantastically rapid implosion of the outer surface of the star at this stage of its life cycle. The gain of energy from this infalling material can be so great as to blow off a small part of the star. This may be seen as a 'nova' explosion, the formation of a new star. Of course this is a complete misnomer; the star may be close to its end.

If it is not much heavier than the sun, the star may adjust gracefully to this new estate of affairs, its interior material being sufficiently robust to stand up to the gravitational attraction between its parts which is causing it to collapse. But a heavier star will have a stronger tendency to condense, and may even compress its atoms together so strongly that their electrons are driven into the nuclei, the resulting electrically neutral objects pushed together to make one big nucleus. The object so formed is termed a neutron star, so-called because it is made up almost entirely of neutrons. These are electrically neutral objects which, along with the positively charged equally massive protons, go to make up the central nucleus in an atom. So the neutron star can be considered the nucleus of some giant atom.

We conclude that if a star is only a little heavier than the sun it becomes gradually fainter as its collapse slowly halts, and it turns into a white and ultimately an invisible black dwarf about the size of the earth but as heavy as the sun. If it is somewhat heavier, the fate of the star is to collapse to a much smaller and far more compressed neutron star about twelve miles across. White dwarfs have certainly been seen in the sky; neutron stars have recently been recognized as pulsars, point sources of very rhythmically pulsed radio waves. In fact these pulses were so accurately spaced that they were at first thought to be signals from a far-off intelligent civilization; they contained absolutely no information however, so this idea had to be rejected. At best they are the signals of morons!

A neutron star is indeed a remarkable object. It is a homogenous ball of neutrons at enormously high density, on average a thousand billion times heavier than water. A matchbox full of its material would weigh a million tons! It is also very small, being no bigger than twelve miles or so across. Nor does it have any mountains that are noticeable; the highest is no taller than eight inches. The interior of the neutron star is expected to be remarkably like that of the earth, with an outer solid crust and a central solid core with a molten interior. Indeed because of the homogeneity of the material of which a neutron star is constructed we can feel more certain about understanding its structure than that of the earth. What a strange world in which we understand more about worlds billions of miles away than that under our feet.

It doesn't appear all that easy for a star to become a neutron star; the most prominent pulsars have each been recognized at the remaining core of a far heavier star which blew itself to pieces in a supernova explosion. Some outer parts of the exploding star may even still be observed streaming out in space from the original place of the explosion, as has been found in the case of the Crab Nebula, the expanding shell of a star which erupted in a supernova in A.D. 1054. That explosion was witnessed and recorded by the Chinese at that time; drawings of it have also been found in prehistoric caves in several places in North America. It undoubtedly made a great impression on earth at the time.

It is by means of a supernova explosion that a massive star can become a neutron star, and so find a peaceful resting state. But there is only one such explosion seen every hundred years in each galaxy, though still enough to make elements heavier than iron in the catastrophe to allow us to have sufficient uranium and other heavy elements here on earth garnered from the shattered remnants of the explosion. But there are far more heavy stars than that still condensing out of interstellar material, so few of them will be able to attain the peace and security of black dwarf or neutron star state.

What happens to those stars which cannot attain a tranquil old age as a neutron or dwarf star? The answer we are faced with is a very disturbing one indeed: it just vanishes! It becomes what is now called a black hole. Yet it still has an effect on its surroundings, rather in the fashion of the invisible man. He cannot be seen, but he can see and hear; in fiction he is usually allowed to speak and knock over objects, so giving away his position. This is very similar to the effects of a black hole, but it turns out that black holes open up a range of experiences which are so fantastic as to be scarcely credible. They are so far beyond anything met in nature up to the present that not even science fiction writers have used them in their flights of fancy.

To appreciate how remarkable black holes are, let us first see why they have the name they do. An object is black if it absorbs all the light falling on it and emits none in return; it is only invisible if it does not affect light at all, but allows light rays to pass through it undisturbed, as in the case of a sheet of clear glass. So we expect a black hole to soak up all the light shone on it – very different behaviour than if it were an invisible star. But why does a collapsed star have this property? That is what we must discover, and it is the answer to this question which will lead to other curious properties, some with enormous implications for man's future.

Suppose we are foolhardy – or brave – enough (according to one's point of view) to try to stay on the surface of a collapsing heavy star. We might do this to observe at first hand what is occurring, or to win a bet, or even because we don't have enough energy to leave when the going is good. As the star collapses it becomes more dense and we approach closer to its centre, and while doing so the gravitational attraction pressing down on the stellar surface increases. This change of gravity with distance away from the centre of a massive body can even be felt on earth; at the top of Mount Everest the force of gravity is one fifth of a per cent less than it is at sea level. As the star collapses the velocity at which we would need to fire a rocket to escape from its surface will gradually increase. Again this is something we have already experienced in man's adventure into space, it being easier to escape from the Moon, with its fourfold lower gravity, than it is from Earth.

Let us imagine that we have a space-ship prepared and ready to leave the star whenever we feel the going is getting too rough. As the star gets smaller and we and the rocket get closer to its centre the escape velocity increases. On the face of it we could leave at any time provided we had enough fuel and a fast enough rocket; we would not feel that we might get trapped in such a situation. But that would be a fatal mistake, since there is a cardinal rule, discovered by Albert Einstein at the turn of this century, that we cannot travel faster than the speed of light. This has been found to apply with extremely high precision to any material thing, and we do not expect it to be violated as our star is collapsing. When it has grown so small that the escape velocity of our space-ship to freedom has reached light velocity we would only just be able to flee; an instant later we would be trapped on the surface of the star for ever. Nor would we be able to send any message about our predicament to bring us help; even a signal sent using light itself would not have enough speed to escape the powerful attraction exerted by the collapsed star on all objects on its surface.

This critical size of a collapsed star, inside which all objects are trapped forever, is called the Schwarschild radius of the star, in honour of the German scientist who first discovered the possibility in 1917. Its size for the earth is only one centimetre; if the earth was compressed to within a sphere of that diameter it would never be able to escape from it again, nor would anything on its surface ever be able to do so. This size is so small, only one billionth of its actual radius, that it is difficult to imagine it being compressed so enormously.

There are situations where such high compression isn't necessary to produce a trapped surface. If a star is extremely massive then the density at which all its material is inside its Schwarschild radius may not be very high. For example a star one hundred billion times as heavy as the sun in such a collapsed state would only need to be as dense as water. It is possible to imagine an advanced civilization who experimentally construct a heavy collapsed mass by re-routing all the stars in a galaxy so that they would all pass through a region a little less than the critical Schwarschild radius (for their total mass) at about the same time. If the galaxy has about one hundred billion stars then the stars would have to be aimed at a region with roughly fifty times the radius of the solar system. This is plenty of space to contain them, with very little difficulty of collisions if they are carefully guided. Once they are inside the Schwarschild radius their mutual gravitational attraction will hold them together from then on. What is more, no light would be able to escape to the outside because of this strong attraction, so the galaxy would suddenly disappear from view!

It is not necessary to go to such heavy objects before black holes are formed, since any collapsing heavy star which cannot shake off most of its mass rapidly enough must condense into one. This inevitability has been shown to follow from the very successful theory of general relativity which Albert Einstein developed in 1915. This idea was basically that since all bodies fall equally fast the gravitational force acting on them can be described in a manner independent of the falling bodies. Gravity is an intrinsic property of space, argued Einstein; the way it is affected by heavy stars embedded in it is that it becomes curved. We notice this curvature when we travel past these stars, since we then move not in straight, but in curved lines around them. Our earth does this around the sun, so we already experience the curvature of space; it is also curved by the earth, and this curvature affects us even more strongly. Einstein showed how this distortion of the space around a heavy body could be found in a very elegant fashion; his rules make up the general theory of relativity.

We can picture curved space most easily if we just consider beings living in a flatland. They have height and breadth, but no thickness. Their world might be an infinitely extensive flat plane; heavy stars in it would cause local crinkling of the plane. The heavier the star the more the plane is distorted, and there is a critical size which would cause the plane to fold over on itself. It could then be considered as the surface of a balloon, and might even become separated completely from the original plane. The flatlanders on the balloon would then live in a closed universe from which they could never communicate with the rest of the world.

This is the sort of separation which occurs when a black hole is formed by a collapsing star. The part of space very close to the star becomes so curved by the enormous gravitational attraction around the star that it becomes completely closed off from the rest of space. Nothing that is inside the closed-off region can ever escape out into the rest of the world. No particles, no observers and not even light can escape, but curves back to stay always near the star. However, this trapped region is still accessible to the unwary traveller from outside. Indeed the outsider is strongly attracted to enter the trap by the enormous gravitational pull exerted by the collapsing star. Thus the separation of inside from outside is only of a one-way nature. That is why the trapped region is called an 'event horizon' – nothing happening inside can ever be communicated to the outside. To the insider his horizon is precisely at the edge of the critical Schwarschild distance; that is the end of his world.

This collapse of a heavy star has been very carefully

analysed using Einstein's theory of gravitation, and no loophole has been found to the formation of an event horizon around it. Of course we could discard the theory, but it would be very unwise to throw away one which has accurately predicted the deflection of light by the sun and the minute deviations of the orbits of the planets from those expected by the three-hundred-year-lasting theory of gravitation of Isaac Newton. Not that we should have unshakable faith in Einstein's theory either; it displaced Newton's, and could be found wanting in its turn.

In fact the formation of event horizons seems to arise in other alternatives to Einstein's theory of gravitation. All that is needed is the description of gravity as an intrinsic distortion of space, together with the existence of a limiting velocity, that of light, for all motion. These together give rise to the existence of one-way membranes. Even in Newton's theory, which Einstein's theory displaced, a black hole can be created, and was predicted as long ago as 1789 by the French scientist Laplace. He pointed out that a sufficiently massive and compressed star should be invisible because the escape velocity from the star's surface would be faster than that of light; no light could escape from it. All in all, the black hole has a very high level of theoretical legitimacy. But in any case, whatever the theoretical situation, we must be prepared for the possible production of event horizons; if they are lurking around out there in space ready to trap us then we must know what sort of properties they might have and especially how to avoid them.

One of the properties which was originally suggested but recently been seen to be wrong was that the event horizon at the critical Schwarschild radius has very nasty properties, such as squashing an observer to zero as he falls through it. That, we know now, should not happen; he feels no bump as he disappears from the outside world never to return. However it has been realized that a very curious situation does arise due to the way that time is distorted near the black hole.

A space traveller will take a finite time to fall into the event horizon around a collapsed star. At least he will think so if he measures the duration of his trip by a watch or clock he carries with him. But according to an observer who keeps a good distance between himself and the disappearing spaceman he will only see the rocket and the rider going ever more slowly to the star, getting fainter and fainter as they do so. It will take the space-craft for ever to reach the critical Schwarschild radius, at least according to the onlooker.

This slowing and fading will also have happened to the collapsing star itself, at least as far as our canny watcher staying outside is concerned. He will think that he is seeing a frozen star since as its surface approaches its Schwarschild radius it will travel ever slower, ultimately appearing to cease to move. That is why the early workers investigating collapsing stars called them frozen.

The slowing down of the collapse can be made very clear if we suppose that the falling space traveller shines a flash of light to the watcher at regular intervals. The unlucky astronaut thinks he is signalling at regular intervals, but as he descends to his fate at the event horizon, his watching colleague will notice that the interval of time between one flash and the next is getting successively longer as the descent appears to slow down and finally be frozen. Ultimately the time between flashes will be so long that it will exceed the watcher's lifetime and even those of his descendants. Yet the falling spaceman is keeping religiously to his regular schedule for signalling.

The sombre title of black hole might well be objected to on this picture, since an outsider can see only a frozen star and frozen spacemen hung motionless over its surface like dead flies on a spider's web. Yet the name is still an apt one for a number of reasons. First of all as the star freezes it gets darker very rapidly, fading from view in four millionths of a second if it is ten times as massive as the sun. So it truly blacks out quickly, doing so with such speed that it would be impossible to spot in the act of 'going out'. Heavier stars vanish from sight in proportionately longer times. Even a star a million times heavier than the sun would disappear in this way within a quarter of a second, again impossible to notice. Only with an object as massive as a normal galaxy – containing about a hundred billion suns – is there a chance of seeing its death, but that still occurs in the brief span of about ten days. That is why the search for imploding stars dying out in this fashion appears impractical.

A black hole truly is black. It is also a very deep (but not quite bottomless) hole as we will now see. For suppose that a mirror was placed on the surface of the collapsing star to reflect back any light shone on it. The distant observer shines his torch at the mirror some time before the star has collapsed anywhere near to its critical event horizon, and has no doubt, by the flash of light returned to him, that the mirror is in good working order. But if he continues to shine the torch-beam at the mirror constantly as the star collapses the reflected beam will suddenly black out as the star becomes frozen exactly at the same time as the star's own light dies sharply away. So it certainly has become a hole in the sense that it doesn't seem to reflect light but only soak it up.

It might be possible to send a further rocket to the frozen surface and snatch up a piece of the star apparently suspended there in space till eternity. Yet that would not be easy since to catch up with the surface of the imploding star it would be necessary at least to fall with it for a short time. As the recovery rocket approached the infalling surface it would have to be very careful indeed to catch the surface before it had gone past the point of no return at its event horizon. However, the crucial difficulty is to be able to send the rocket down fast enough after the collapsing star had been seen to be frozen and blackened in space. If it has been sent down before that then the project is not one of trying to dig a piece out of a real black hole.

A star which has disappeared, apparently for ever, into its event horizon still has a faint chance to reappear if some new source of energy inside the star were available to halt the collapse before the star had actually fallen through its event horizon. But that is cheating, because it is undoing the black hole, and making it light again. Even keeping to the rules, and not allowing new sources of energy to undo the collapse before it is too late, it is still theoretically possible for the space-craft, if it acts soon enough, to land carefully on the infalling surface of the collapsing star, chip a bit off very quickly and shoot away again without being doomed. Such a manoeuvre would need a very fast and powerful rocket (and a cool-headed crew) to achieve; the high speed and power effectively allow the black hole to be speeded up and illuminated, as viewed by the spacecraft, so as to lose its frozen and black character. The star certainly regains its black-hole nature as the space-ship flees from the perilously infalling surface. So the black hole is not quite a perfect hole to the outsider as all that. Yet it truly is if he allows himself to fall inside the event horizon.

It is only by the use of a powerful motor that a spacecraft could hope to escape from the clutches of a black hole. A disabled ship, or a helpless spaceman who had fallen out of his ship, would have a much more unpleasant time. Their fate is almost certainly one of being captured by the black hole unless they are initially moving so that they would miss the centre of the black hole by a suitable margin. A rocket moving towards the black hole at one hundred thousand miles an hour some distance away from it would have to be aiming at a point at least one hundred thousand times distant from its centre than is its event horizon to avoid being captured by it. As the speed of the rocket or man decreases so the possibility of being captured increases, while if they are moving faster they have more chance of escaping the black hole. Yet they are then less manoeuvrable, so that such a gain will be illusory.

The moral from all this would appear to be to stay well away from black holes. Yet they possess such remarkable properties of distorting space and time near them that they could well exercise a fatal fascination over humanity in the future. The particular way in which time is slowed down on the edge of a black hole is especially interesting, so leading the way to a possible time machine. It would consign near immortality to those venturesome enough to travel close to the event horizon, always staying on the safe side. Not that they would feel older in themselves; their life span would only have been lengthened with respect to their sit-at-home friends and relations.

It is interesting to compare this method of keeping young

with that achieved by very fast space travel. If a spaceman left the earth aboard a rocket travelling at 99 per cent of the speed of light he would age seven times more slowly than his colleagues and his loved ones he left behind him on earth. At the fantastic speed of only one ten-thousandth of a per cent slower than the speed of light he would grow old millions of times more slowly. He could find on his return to earth, after visiting a far-off star at such speeds, that his year-old son he left behind on earth was now older than he; at the faster speed he might even return after a long journey to find that the human race had died out.

This remarkable property of time has been shown to be true by various experiments in the past few years. Of course it has not been possible to speed a human being up to such high velocity, and only the very light elementary particles have been observed under such extreme conditions of motion. For example, very light particles called mu-mesons are created at the top of the earth's atmosphere by collisions between energetic protons coming from outer space (possibly from supernova explosions in the galaxy) with air molecules. These mu-mesons are expected to decay into other particles before they reach sea-level. Due to their high speed as they travel down to sea-level the time they experience is about seven times slower than expected by a watcher stationary on earth. Thus they age much slower than if they were at rest; in fact they live seven times longer. This explains why many more of these mu-mesons reach sea-level than should do so without this time dilation effect, as it is called.

Such slowing of time for moving clocks has been tested at even faster speeds; electrons have been sent so fast that a clock carried with them would go about fifty times slower than when at rest. And it has been observed to do so by measurement of suitable subtle effects. Other particles – protons, neutrons and many others – have also been accelerated to similar speeds, and always the expected time dilation has been observed. An even more precise experiment has been performed to silence the doubting Thomases who claim that the effect of time dilation would not actually be experienced by human beings in rapid motion. Two American physicists,

Joseph Hafele and Richard Keating, took a set of incredibly accurate atomic clocks for a round-the-world trip on a Jumbo jet. On return to Washington, the moving clocks were found to be exactly the amount expected behind other atomic time-pieces they had left behind; the actual time-delay was only about one hundred billionths of a second, but still just enough to be measured by their incredibly accurate apparatus.

This slowing of time by motion has produced a paradox, known as the 'twin paradox'. It arises when one of a pair of identical twins makes a fast space trip, returning to his stationary twin at a later date. The latter has seen the moving twin speed off to distant stars and return much younger than he himself. He explains this, as he sees his twin still possessing all the vigour of youth, by saying sadly to himself 'moving clocks go slow'. But the moving twin can just as well see the twin he left behind him on earth disappear from his view, and then reappear at the end of his journey. To the space-twin the earth-twin is also moving; why, then, does he have a long white beard and rheumatism? for surely 'moving clocks go slow', and the earth-twin is moving, as far as the space-twin is concerned.

According to this argument each twin should have aged less than the other, certainly an impossible situation. This twin paradox is resolved by noticing that the situation is not the same for both twins; the space-travelling one is accelerated and de-accelerated at various times, while the earth-bound one is not. It is this difference which causes the moving twin to be younger than his stationary brother when he returns to earth, as was fully realized by Einstein when he discussed this situation and formulated the Theory of Relativity in 1905. It is this theory which shows that 'moving clocks go slow', which has been so amply validated by countless experiments.

We can even use Relativity, with a little extension, to understand why clocks go slow at the edge of a black hole. Consider a man in a lift at the top of the Empire State Building. If the lift suddenly dropped he would fall freely, accelerating as he approached his death on the ground. If he could not look out of the lift he would have no way of telling that he actually was falling; he would be in a free-fall state, and could float gently around in the lift. But if a suitable acceleration can remove a gravitational force then it is equivalent to it. So a space-man enclosed in a rocket-ship near the edge of a black hole would not be able to decide if the enormous force pushing down on him was due to a black hole or was because he was being accelerated upward very fast. Because of this effective acceleration when he has made a round trip to a black hole and safely back to earth, not necessarily at high speed, he is in the position of the travelling twin; he will have aged less than those left behind on earth. The closer he goes to the edge of the black hole the larger is the effective acceleration he undergoes; as he falls towards and later escapes from it, so the time lag is longer.

Such 'longevity trips' would appear feasible, if costly, though a necessary pre-condition for them is to find a black hole first. That would appear to be easier said than done, rather like finding a needle in a haystack. The collapse of a massive star would take place literally in the twinkling of an eye, as we saw before, so it is no good searching the skies for the blacking out of stars, nor even of galaxies. Nor do we expect to be able to see one directly, because once a star has collapsed it is so black as to be impossible to see. There are curious effects which arise if it were possible to look at a black hole illuminated from the other side, or even from the same side. In particular there would be a halo around the object due to light being scattered by it straight back to us, a similar effect to that seen by looking at the brilliant illumination that surrounds the shadow of one's plane upon the clouds, seen far below. In this case, though, it is only innocuous water-drops which deflect the light straight back, this back-scattered light being best seen adjacent to the plane's shadow. But to see the rather faint halo round a black hole we first have to catch the black hole, and that has already been seen to be difficult.

To be effective we need to understand more about black holes, and in particular what special indicators they might have to signal their existence. We also need to understand how a black hole swallows up unwary objects which get in its way.

That has got obvious relevance to the possible ways we can see it, since the objects might emit some sort of radiation as they are destroyed. We also have to consider if a black hole can have properties other than its mass – for example can it have a charge, and if so how much? This could be very important for black hole technology, since it is easier to guide a charged black hole around by electrical forces than by gravitational ones, because the former are so much stronger than the latter. We turn to these questions in the next chapter.

5 Cannibals at Large

A black hole is a cannibal, swallowing up everything that gets in its way. Once engorged by it there is no hope of escape; our own world is left behind for ever on passing into its event horizon. Yet to an observer outside, keeping a careful watch on the heavens, such an inexorable fate can never be observed. An unwary space voyager ensnared by a black hole will never be seen to enter it, but only to become frozen to its surface. If account is to be taken of the discreet particle-like character of light then the frozen space voyager will vanish totally from the outsider's sight within a fraction of a second, though he is still hanging there motionless if he could be seen. This is because from then on the total amount of light he could emit is below the energy of the smallest packet of light the laws of physics would allow him to give out.

Even though it appears frozen when very distant from us, a black hole soon drops its icy stillness as it gets nearer. It exerts a strong gravitational attraction on anything near it, so that we would have to prevent ourselves from being eaten alive by it. The chance of the sun or even the earth suffering such a fate by lying directly in the path of a marauding black hole is expected to be about as small as the probability of a near-collision with a wandering star, and that is very small indeed. But it could happen so we should have some idea of what signs to watch for to warn us of our danger. Such collisions could occur far more frequently than the simple guess we just made because there might be far more black holes around us than we realize. In any case, the old saying goes 'it's better to be sure than sorry', and though mankind has not taken much heed of this in the past we must shake ourselves out of our lethargy and complacency and seriously consider what should be done in the face of the ultimate possible annihilation.

We must also realize that mankind is on the verge of travelling outside the solar system. That might appear to be a rather premature statement in the face of our having got to the moon only four years ago and so far only sent unmanned space vehicles to other planets within the last eight years. But the rate of scientific research is ever increasing and one can envisage at least unmanned space-probes being sent to the nearer stars at the turn of the century. It could be disastrous to have our first attempt at interstellar flight end in the complete disappearance of the ship, so we must try to chart the positions of hungry black holes as carefully as possible.

Seeing a black hole form is almost impossible, unless it is truly of astronomical dimensions. Even one of the size of a collapsed galaxy, containing over a hundred billion stars, would disappear in only a few days. Nor is it possible to see it in the usual manner by observing light shone on its central region. But that leads us to ask if a black hole can ever leave any useful evidence behind it that it was there at all. If there is nothing at all then we could be in for a very risky future indeed, with no hint of the death traps lying in wait for us in the heavens or even heading straight for us. We would just have to cross our fingers and hope luck was on our side.

The situation is not quite as bad as all that. A black hole leaves its signature behind, frozen into the curvature of the space outside its event horizon. This warping of space can be seen in the way the orbits of space-craft or other stars are affected, as is the path of a stone thrown into the air changed by the earth's gravity. If the earth were removed immediately the stone were thrown it would move in a straight line, instead of being pulled back to earth by the earth's gravitational attraction. But if the earth were suddenly annihilated yet no disturbance made in the space around it, then the stone would curve back to the now-vanished earth's surface. The undisturbed space around the earth would appear curved just like that around a black hole, though of course certainly not as much.

Suppose we were tracking the path of a space-ship in distant space and noticed that its path was being deflected, though we could not see any star or cloud of matter nearby which could have caused the bending. We can assume that the ship's propulsion unit is functioning perfectly, so that the only cause of this peculiar behaviour that we can think of is that the ship is moving near a black hole. Can we work out what sort of black hole is causing this disturbance? We would be able to do that but only because it is now fairly well understood that there are only a very restricted range of black holes which can be created in the collapse of a massive star.

The history of the creation of a black hole from a collapsing massive star is one of implosion with the creation of the event horizon. When the nuclear fuel in the interior of the star is all used up, the star collapses in on itself at an enormous rate. It then falls inside its own event horizon within a fraction of a second. Before the implosion the star may have been rotating, and it is expected that this rotation will even speed up as collapse proceeds. This is exactly what happens when a skater is spinning round slowly with outstretched arms and twirls faster as he brings his arms into his body. This conservation of spin could cause an important proportion of the matter of the star to be spewed off into space, but calculations seem to show that there isn't always enough mass lost to allow the star to age gracefully as a dwarf or neutron star. So for the unlucky stars the collapse then proceeds, and a black hole is formed.

Even if the star initially had a lot of mountains on its surface they apparently leave no trace behind them when the event horizon has been formed and the star has vanished behind it. The frozen warp of space left behind as the black hole's signature appears the same as if the collapsing star had been as smooth as a billiard ball. That is why it can be said colloquially that 'a black hole has no hair'. It is only possible to recognize, in the frozen curvature outside the event horizon that an object with a certain total mass, total spin (and also total charge) had caused this distortion of space. All the other distinctive features of the star which would allow it to be distinguished from any other star, such as the total number of its neutrons and protons, or its chemical composition, are all lost to the outsider. Only if he dares to venture inside the event horizon can he ever discover what it was that had originally collapsed (though he couldn't do very much with the knowledge).

Such a disappearance of matter in the formation of a black hole is quite remarkable, violating as it does some of the most sacrosanct laws of the natural world. One of these, in particular, is worth noting here, because it affects our own individual continued existence. It is concerned with the disappearance of protons and neutrons in the black hole. These two particles are usually grouped together and called a nucleon since they have such similar properties when they are inside the nucleus of an atom. This nucleonic particle – proton or neutron – is fantastically stable, having an observed lifetime over a billion times longer than the life of our sun. This is a very good thing for us, in fact, since if nucleons did decay with any noticeable rapidity then the atoms of which we are all constructed would themselves change their character. This could be disastrous for us if it took place in any amount within a lifetime.

Here on earth we are reasonably certain that the nuclei of our atoms preserve themselves very well against decay. But if they are once inside a black hole they cease to possess their particular nucleonic character to those outside the event horizon. Gravitational collapse destroys the conservation of the total number of nucleons in the world. There we have one of the cornerstones of our understanding of the world about us swallowed up in the black hole. But then, we must be prepared for many differences between our own safe world and that behind the horizon.

Outside the collapsed star the static distortion of space will change the motion of a space-ship or planet moving nearby in a very characteristic manner. If the track of the object is plotted carefully it would be possible to work out the mass and spin of the black hole; to find its electric charge we would have to observe the motion of a charged particle. It turns out that a black hole which is rotating has a very important difference from one at rest. The event horizon still exists, as we would expect, though it is now smaller than in the stationary case by an amount depending on the spin of the black hole. But as the rotating star collapses it becomes frozen, as seen by an outside observer, before it reaches the event horizon. The region on which the star's surface appears to hover for ever has been called the ergosphere. It is the surface on which time stands still, and immortality is created.

Here we have the first glimpse of the time machine of the future. Living on the edge of the ergosphere has none of the dangers that are associated with event horizon brinkmanship, but yet all the advantages in the control of time. If the collapsed star is heavy enough the distance between parts of the ergosphere and the horizon can become considerable, so the danger of falling further down into the event horizon is minimized. So the ergosphere of a large rotating black hole is the place to stay for a while if you wish to travel a few thousand (or million) years into the future. Naturally the longer the journey desired in time the closer to the precise surface of the ergosphere it is necessary to go. The harder it is then to escape back to the sane world of low gravitational attraction from the attraction of the black hole. The cost will thus be higher the further into the future it is wished to travel, but that is to be expected. One never gets something for nothing!

Let us return to our quest for black holes in space. The two methods of identification we have come up with so far – watching the star 'go out' as it collapses to a black hole, or observing how space-craft are deflected by otherwise undetectable matter nearby – are neither very propitious. Both require one to have a very good guess as to where and when a black hole is about to form or its whereabouts, after it has formed. But that is exactly what we are trying to do in the first place, so we have to consider more evident signs of black holes in the heavens.

The most obvious thing to look for is the possible effects black holes distributed throughout space would have on the surrounding stars. If there were many black holes then stars would move in an incomprehensible fashion, jostled this way and that by the nearby marauders, apparently without rhyme or reason to distant watchers who are ignorant of the existence of black holes. Indeed this has been the case for several decades; there appears to be too much give and take between the stars in several galaxies to be accounted for by the observable stars in them. This phenomenon has been especially observed in galaxies which have an elliptical or Rugby Football shape. The ratio of the actual mass they contain to the light they emit has been measured for a number of them and found to be about sixty or so times greater than that of the sun; as for example, in the case of the elliptic galaxy NGC4486. Our sun is an average star in our galaxy and appears to be similarly normal in elliptic galaxies. Nor do we expect there to be much invisible normal matter in these galaxies since they are particularly devoid of clouds of gas or dust.

This raises the important question as to what the invisible matter is in elliptic galaxies. It would seem that a galaxy such as NGC4486 may contain as much as 98 per cent of its matter in a form we cannot see. It is natural to suggest that it occurs in the form of black holes, obviously invisible to us except by the extra energy they impart to the neighbouring visible stars in their galaxy.

It is possible that there are quite large collapsed objects at the centres of certain elliptical galaxies, formed by settling of gas, shed by evolving stars, to the centre of these galaxies. There is no optical evidence of any especially black region at the centre of such galaxies, though there are some interesting cases of giant elliptical galaxies in which rather curious events are occurring. The second brightest elliptical galaxy in the Virgo cluster, M87, is an X-ray and radio source, one component of which is very small and could coincide with the optical nucleus. It has been suggested that the X-rays are emitted by hot gas shed by stars falling into a collapsed object at the centre of M87. If so, then M87 is committing suicide. How many other galaxies are doing the same?

A similar puzzle exists for invisible galaxies in galactic clusters. These dark members of a cluster should exert visible effects in their companions in proportion to the total number of them, very similar to the effect of 'black stars' on their companion stars in a galaxy. Various clusters of galaxies have again been found to have a high proportion of this dark matter, usually between ten and a hundred times more than that seen. A case in point is the Virgo cluster containing seventy-three galaxies and with about fifty times more invisible than visible matter. Here we may be faced with a far more serious problem, since the missing matter may have to be considered as collapsed galaxies. Not only do we find that some galaxies may be mainly composed of black holes but that most galaxies themselves are black ones. We may be able to avoid that latter situation if the observed galaxies in the Virgo and in similar clusters contain all the missing mass, but that seems rather unlikely.

A way out of this rather unpleasant situation, that the normal matter we live on is so rare as to be regarded as abnormal, is that there are just far more dim stars around than we had realized up to now. However, strong support for this enormous amount of invisible matter to be actually occurring as black holes comes from a careful analysis of the abundances of heavy elements in the stars. Those stars which are some distance away from the plane of our galaxy are particularly relevant in this respect, since they contain surprisingly large proportions of elements heavier than iron. The only way they could have acquired these elements was by soaking them up from the surrounding interstellar clouds.

Heavy elements are not expected to have existed in any quantity in very ancient times in the Universe. They are thought to have been created in super-nova explosions of massive stars at the end point of their nuclear burning; as time goes on more heavy elements are being formed in this way. This model of heavy element formation meets a difficulty when trying to consider the abundancies in stars out of the galactic plane because these stars are considerably older than ones in the disc of the galaxy. To account for this anomaly it is necessary to assume that there were a suitable number of supernovae explosions about ten billion years ago. These, it turns out, could best have come from a large proportion of very massive stars with rather short lifetimes. But at least some of these stars will be so massive that they will find it impossible to avoid the ultimate fate of any body in the Universe – to collapse and turn into a black hole. So we expect these black holes also to be mainly out of the galactic disc but lie in the spherical halo on each side of it.

Calculations made along the lines of the above ideas suggest that about 90 per cent of the stars in the Universe are of the type that can become black holes. This is somewhat lower than the values of about 98 per cent of all matter having collapsed away, but is quite consistent with this higher value if we remember that a reasonable proportion of the invisible matter in galaxies or clusters of galaxies could just possibly be in the form of normal invisible matter, such as clouds of dust or gas or small stars. But that still makes the amount of black holes and black galaxies uncomfortably high.

We have to reduce this estimate for our own galaxy, since the amount of invisible matter it may contain cannot be more than 10 per cent, as calculations have shown. This would agree with the estimate of seven stars born every year of mass larger than the critical value for collapse (about twice as heavy as the sun), if only about one-seventh of those once created could not lose enough mass during their lifetime to prevent turning into black holes. That such a relatively small fraction of stars ultimately become collapsars has recently been borne out by careful dynamical calculations.

So far, however, we have only used rather indirect evidence. Is it possible to think of a way of directly 'seeing' a black hole in the act of creation? We would have to use different methods than looking for it with visible light since a black hole would 'wink out' too quickly to be noticed. But we could ask what sort of radiation might be produced in such a cataclysm which would act as a black hole's signature tune.

When a star explodes in a super-nova it emits an enormous amount of visible light. Is there radiation of any sort which could be emitted in large amounts when a massive star collapses? Let us consider how we would experience here on earth the sudden complete annihilation of the sun. Disregarding the problem of the earth's rapid cooling we would notice that it had stopped moving in an elliptical orbit but was speeding off nearly in a straight line after the removal of the sun's gravitational attraction. We would not expect this to happen immediately but occur at the same time that the sun vanished from sight. In other words we would still be orbiting round the sun for about four minutes after it had actually been annihilated. It takes that period of time for a modification of the sun's gravitational pull to reach us, and we can regard this spread of gravitation as itself a form of radiation, which we can naturally enough call gravitational radiation.

The radiation we call light has been observed ever since animals acquired visual receptors, but no animal has found it of any survival value to be able to detect gravitational radiation. That is because there is very little of it around, or if there is it has very slow variation in any reasonable amount on earth. But this is exactly what would be expected; gravity is a very small force, and it is only when bodies are astronomical in size that the gravitational force between them becomes noticeable. In order to observe gravitational radiation we need either to have a very large detector, say the size of the earth, or a specially built, very sensitive detector. On the other side of the coin we must only expect to look for gravitational radiation from processes involving very large masses.

But that is exactly where we came in. The collapse of a massive star to form a black hole would be exactly the sort of cataclysm which might produce detectable quantities of gravitational radiation. Not that we are certain that it is reaching us in large enough quantities to be detectable nor do we know in which direction to look. The only thing to do is set up as sensitive an apparatus as possible without very much directional sensitivity, and then switch on and keep our fingers crossed. And that is precisely what Joseph Weber did at the University of Maryland, starting in about 1969.

The results of these experiments were quite startling. Weber used aluminium cylinders suspended by wires in a vacuum, each cylinder being about a metre and a half long and a metre across. The cylinders were expected to be excited by any incoming waves so as to be set vibrating along their length. It was these oscillations which had to be detected, and this was done by quartz crystals bonded to the surfaces of the cylinders. The sensitivity of the system was so high that displacement of as small as a thousandth of a nuclear diameter could be measured.

Because of this high sensitivity it proved easy to pick up

all sorts of uninteresting side-effects, such as demonstrating students marching around the campus. To avoid being confused by such demonstrations another detector was set up about a thousand kilometres from the first, near Chicago. Only oscillations detected at the same time by the two separate detectors were regarded as arising from an extra-terrestrial source of gravitational radiation.

Weber announced in 1969 that he had observed several hundred disturbances over several months which could not be explained as chance fluctuations. It was found that these signals were most frequent when the detectors were pointed so as to be most sensitive to radiation coming from the centre of our galaxy. The variation of signal intensity was very close to twelve hours, so it seemed as if the radiation came through the earth with very little disturbance and was picked up by the detectors when they were pointing at the galactic centre but were even on the other side of the earth to it.

The most exciting feature of the radiation was that it consisted of a short pulse of less than half a second long about once every four days, and was picked up at a frequency of about 1,600 cycles per second. Brief outbursts of radiation of such frequency contain a considerable amount of energy and if they are produced from the centre of the galaxy would correspond to it emitting up to two hundred solar masses as gravitational radiation each year. That is an enormous amount of radiation, and we are left with the disturbing problem of explaining where it came from. It certainly would not be satisfactorily explained by the final collapse of a star which had imploded as a supernova, for that happens only once every hundred years or so in our galaxy; Weber saw it happening once every four days.

This puzzle has disturbed scientists so much that some of them have even been prepared to throw away the geometrical theory of gravitation on which a lot of the interpretation of the evidence is based. Yet the experiment has not yet been repeated by other scientists, so it is still to be regarded with some caution. Hopes of using the earth as a far more sensitive detector have been dashed by the large amount of seismic noise it contains, whilst apparatus has only just been set up on the moon to observe its vibrations in the same fashion, no data being yet available.

A suggestion has been put forward recently that explains the evidence as it stands but may well send shivers up and down a reader's spine. The radiation, so it is proposed, is coming from stars near the galactic centre as they fall into a large highly spinning black hole which forms the central part of our galaxy. The mass of this terrible nucleus could be as high as one hundred million times that of our sun, and it is gobbling up the stars on its outer rim at between 1 and 30 solar masses each year. In this picture we are lucky to be in the disc of the galaxy since if we were a little to one side or the other of it the radiation would be nearly invisible. We would then have very little evidence of the central black hole, at least for a long time to come.

It is the decided difference between the amount of energy radiated in the plan of the galaxy and that in other directions, caused by the spin of the black hole, which allows the model to work, for if it radiated at the same rate in all directions Weber's results lead to a mass loss of nearly a thousand suns each year. Over a period of a billion years, about one-hundredth of the age of our galaxy, such a loss would have produced disturbances which we should now be able to see; there is no evidence of such effects. This strongly suggests that only a highly spinning super-massive black hole at the centre of our own galaxy is causing the emission of this radiation. We truly seem to be in a dangerous world, what with this trap at the galactic centre, let alone the black holes possibly making up a preponderant bulk of the matter throughout the rest of it.

There is certainly evidence that violent events are occurring at the galactic centre. For example an arm-like structure, mainly of atomic hydrogen about nine light-years from the centre, can be seen approaching us at a speed of about thirty miles per second. This motion is seen by the change it effects in the wavelength of radio waves emitted by the hydrogen, especially at twenty-one centimetres. This and other radio evidence certainly support the suggestion that the galactic centre is full of activity, and even of harbouring a black hole.

It is possible to obtain evidence on black holes a little closer to home by investigating systems composed of two stars. It is thought that a tenth or more of all stars are in pairs, but in some cases only one of them is directly visible, the other being too small. It is possible to estimate the mass of the invisible companion in some of these cases, and values higher than that allowed for stable neutrons or white dwarfs have been found. If they are black holes it is expected that in the formation they ejected a considerable proportion of their own mass. This would cause the motion of the two stars round each other to become considerably disturbed, and so give rise to what are called eccentric orbits. And indeed some evidence of this eccentricity has been found recently in binaries with one invisible companion.

An interesting example of this is in the case of Σ-Aurigae. The visible star is eclipsed periodically, apparently by an invisible companion. The black component is most likely over twenty times heavier than the sun and a little lighter than the primary star. The secondary component apparently now possesses a disc of dust particles circling round it which are emitting infra-red radiation (possibly as they fall into it), and this has been observed. At the same time the details of the eclipse are neatly explained by the rotation of the disc round the visible star. Here then we have a possible black hole in our galaxy; by some it is claimed as a very good candidate.

There is an even better candidate discovered recently which is the X-ray star Cygnus X – 1, about six thousand light years away from us. It emits X-rays, which are very energetic beams of electromagnetic radiation, in such amounts that they are visible on earth. At least they would be if the earth had no atmosphere, but they are almost completely absorbed by it. They have been detected only in the last year by means of equipment mounted on the Uhuru space satellite. These X-radiations have the very peculiar property of varying remarkably in their period from one-tenth to ten seconds over a duration of days. Very recently the X-ray source was identified with a visible blue supergiant star. Careful observation has shown this star's speed to be varying in a sine-wave pattern, with a five and a half-day period. This indicates that the visible star must have an invisible companion dancing around it, which calculation shows must be at least three and one half times more massive than the sun; the visible primary should weigh about fifteen solar masses if it's a typical blue star. Further data on the invisible secondary can be obtained from changes in the blue light emitted by hot helium atoms in the system. This light is most likely formed by its companion sucking away chunks of the star's outer atmosphere. This is in agreement with the fact that light is found to vary at a different rate from that of the primary, so arises from helium gas heated up around the secondary. The invisible companion must then have a mass of ten solar masses to explain the phenomenon.

The crucial problem is to understand the rapid variation of the X-rays that are being emitted by Cygnus X – 1. They can only arise from a system no more than one tenth of a light second across, or about twenty thousand miles, about the size of a white dwarf. There is, however, no known mechanism to produce such X-rays from a white dwarf, and the only object which can achieve this emission, being at the same time as massive and compact as it is, must be a collapsar. In other words the invisible companion to the blue supergiant star in Cygnus X – 1 must be a black hole. This evidence is still not perfect, since the evolution of blue stars in binary systems is not as well known as it should be, but it is still impressive.

There are also other X-ray sources, such as Scorpio X – 1, which may be black holes; more careful investigation is needed before their secrets are revealed.

The final piece of evidence for black holes indicated that they could be involved in the most fantastic energy sources in the Universe. These are the quasi-stellar sources, or quasars for short. They are so-called because they appear to have no internal structure even when observed with the largest available telescopes. They must also be remarkably compact, since their luminosity varies considerably in a coherent fashion within a period of a year or so. For this to happen the size of the quasar cannot be much larger than several light years since the whole of the quasar must change at about the same time to preserve the coherency of the emitted light.

There is quite good evidence that some of the quasars are

the most distant objects visible in the heavens. They are thought to be speeding away from us at a very high rate, as can be seen by observing how much energy is lost by their emitted light in actually getting away from the rapidly receding quasar surface. This effect is called the Doppler shift, after the scientist who realized that it is a shift of the frequency of the emitted light, and can be used to measure the speed of the fleeing object. Distant galaxies have all been found to have a speed of recession which is proportional to their distance away from us. This rule can be used to measure the distances of very far galaxies, and when applied to quasars puts them on the edge of the observable universe.

If the quasars were stars in our galaxy they would be emitting a reasonable amount of radiant energy. But since they are so very much further away their energy output must be enormous. They are also very small and so created a very great puzzle for the scientists. Here again the black hole steps in to the rescue, since it is compact and if rotating rapidly enough it can capture material and spew out half of it as visible or ultra-violet light and other forms of energy. This has been seriously proposed as a model of the quasars, and describes them as galaxies towards the end of their lives. Their centres have collapsed into compact black holes which are now sucking in the remaining stars around their rim. It would seem that they could still go on doing this for further hundreds of millions of years, so this mechanism of energy production in quasars is very satisfactory.

If this is the fate of our own galaxy then it is not so satisfactory. There is already evidence that there is a black hole at the centre of the galaxy, and as time goes by it can only get bigger as it swallows up all that is in its vicinity. In fact, once a single black hole has been formed anywhere the fate of the rest of the world is sealed: unless it can flee rapidly enough all matter will ultimately be dragged into the hole or end up as gravitational radiation. Such a fate may take a long time to arrive if the black hole is in another galaxy, but the engorgement of our own galaxy by its central black hole may not be so lengthy. Indeed if quasars can be formed in the numbers they appear to have been then a lifetime of tens of billions of years may be all that is left to the Milky Way. That appears an enormous length of time, but we expect to have to flee our galaxy by then to seek safety elsewhere in the Universe. And it may even happen far sooner.

None of the evidence presented in this chapter is conclusive because we cannot actually go out and see if Cygnus X – 1 actually is a black hole or if a particular quasar is a gigantic one. But it all fits together very well, and should cause us to take black holes seriously. They obviously pose a threat to us of a very direct physical nature. Yet that threat, serious though it could be if we were confronted by a cannibalistic black hole on the rampage, is in no way as fearsome as the one presented on the psychological level. For the existence of even a single black hole in the Universe threatens our very concepts of time, of space, of immortality, and of all the myriad commonsense ideas we have about the way the world works. The black hole makes all of these notions appear completely outmoded. A world in which time can vary almost at will, space can curve back on itself and the Universe can be full of traps for the unwary space traveller is certainly a strange and threatening one to our earth-sheltered minds. We have to be prepared to face these changes without flinching.

Possibly they have already faced mankind. Perhaps in the past travellers from far-off stars have conquered the black hole and harnessed its power to drive through the heavens to visit us here on earth. Have records of the past described these visitors and their strange craft? We might even now react with awe if such a space-ship visited our planet. Hopefully that is all we would do and not resort to violence to grasp the immeasurable power it contained from its controllers' hands. But who knows?

6 Taming Black Holes

Black holes are dangerous. Once swallowed up by one, it would be impossible to escape from it. If there are any in the Universe, and we've seen that there is a high probability of this being so, even in enormous numbers, they should be avoided like the plague. Yet they involve matter compressed to the highest imaginable degree, and even beyond. In the collapse of a star to form a black hole vast amounts of energy will be liberated, and while matter is being eaten up by an already formed black hole further energy should be produced. So we expect the black hole to be a source of great supplies of energy.

Energy is the motive power of the Universe. Its nature lies at the heart of the mystery of our existence as animate beings in an inanimate Universe. As William Blake wrote nearly two centuries ago:

> 'Energy is the only life and is
>   from the Body;
> and reason is the bond or outward
>   circumference of Energy.
> Energy is Eternal Delight.'

In the hands of intelligent beings, such as ourselves, a large enough supply of energy will allow us to 'conquer the world'. So black holes have the attractive possibility of helping us towards such a goal. And if we don't have enough energy we may well destroy ourselves fighting over who is to have the little there is. The problem facing us then, is of whether it is better to risk a fate worse than death in the unbreakable embrace of an event horizon, or whether to be prepared for extinction because of our inability to keep up with mankind's voracious appetite for energy.

There are good reasons why we must take black holes as a possible energy source seriously. Mankind is rapidly running out of chemical fuels on earth. Even ignoring the enormous increase of population bringing increasing demands, man will have exhausted the supply of fossil fuels on earth in less than a century. He cannot use fission fuels as replacements so effectively without befouling his globe with highly poisonous radioactive wastes. This problem should be removed (at least partially) by the development of a controlled fusion reactor, though there will undoubtedly still be problems of disposal of contaminated by-products.

In any case, all of these methods still involve energy production by a process with less than 1 per cent efficiency. In the case of gravity there is no apparent reason why we cannot have a much higher efficiency in power production – even approaching 100 per cent.

The energy we expect to obtain from a black hole is gravitational. It arises by gravitational contraction in the same way as does the flow of water from a reservoir to a turbine situated closer to the centre of the earth. The collapse of a star produces an incomparably larger amount of energy than that, but it may be possible to control rapid emission in the black hole case so as to make it a useful source of energy. This problem is in some senses equivalent to that of controlling the process of nuclear fusion so that energy generation is at a usable rate. However we have on our hands another and even more difficult question, that of rendering the black hole innocuous. We have to be able to tame it so that energy generation can be obtained from it in a continuous fashion without constant fear of falling inside the event horizon.

It would seem that gravitational energy is the best form to obtain because it allows a higher proportion of itself to be used than do other forms of energy. In any process of energy transformation the energy is degraded into that involving less order and more randomness. A good example of this is when cream is poured into a cup of coffee and stirred; it mixes thoroughly with the coffee, and never returns to being in a compact mass. Similarly cigarette smoke exhaled from the lungs spreads throughout a room instead of curling back to the smoker. In both cases the end product is a system with less order than it began with.

It is usual to consider the converse of order, or the amount of disorder involved in any form of energy; this is called the entropy of the system. There is a fundamental law of the material world, called the second law of thermodynamics, which requires that no process can reduce the total amount of entropy or disorder in a closed system. Ultimately the system will reach its equilibrium state of total disorder and maximum entropy; it will then have died what is picturesquely called a 'heat death', with all its parts having equal amounts of energy. Nothing new can happen in such a corpse. A death of this type has been envisaged for the Universe, and was described imaginatively by Olaf Stapledon 'Presently nothing was left in the whole cosmos but darkness and the dark whiffs of dust that once were galaxies.'

Gravitational energy has far less entropy than that involved in the nucleus or in the thermal motion of atoms. It is expected to have no disorder at all as far as normal gravitational fields are concerned. But the presence of black holes will introduce a loss of information as they swallow up normal objects, and so introduce a type of disorder. This might prevent us from obtaining the complete transformation of gravitational into other forms of energy by using them; this is a further question we have to consider about black holes. We certainly expect that a black hole will possess some form of entropy, otherwise we could just bring about a loss of entropy by throwing a very disordered system into a black hole, and so effectively destroying the system along with its disorder. This would contradict the second law, that entropy can never decrease; it is like all such laws in that it is a very useful bookkeeping device, so we will attempt to preserve it. To do that we will have to introduce the notion of the entropy of a black hole; this will, in fact, prove invaluable in calculating how much energy is available in a black hole.

To see what a black hole can do for us, and what we can

do with it in return, we have to go back to the basic properties that such an object possesses. Let us start with a collapsar which is non-rotating and electrically neutral. All it possesses, as seen from the outside, is a massive condensed star hidden behind a spherical event horizon. Nothing can escape from this horizon, so it is a one-way membrane. It is this latter surface which must completely characterize for us the collapsed star hiding behind it. That the horizon is spherical tells us that the black hole is not rotating. It is the total surface area of the membrane that is an indication of the total amount of matter it contains. As the amount of matter in the collapsar increases so the surface area increases, actually as the square of the mass. The area is quite small for a reasonably sized collapsar; for example, one ten times as massive as the sun has only a total surface area of about five thousand square miles, the size of a region seventy miles across and seventy miles wide.

You might expect that once such a one-way membrane has been formed it remains the same for ever. Certainly from the outside the frozen character of the collapsed star's appearance would support this. But a membrane can expand, and this one certainly will do so if it absorbs further particles, such as interstellar dust or even neighbouring stars; these latter could cause a drastic increase in the black-hole size. It is more difficult to work out what will happen if two or more black holes collide. While the detailed changes which should occur are far from clear, one thing is certain: the total area of the one-way membranes after the collision cannot be less than that before the collision. So if two black holes coalesce then the sum of the areas of their separate event horizons, which won't necessarily remain spherical as they approach each other, is not greater than the area of the final one-way membrane around the object formed by the coalescence.

It seems here that we have in the surface area of the one-way membrane, a good candidate for the entropy or degree of disorder of a black hole. The total entropy of a system of black holes can never decrease; if we add this entropy to that of normal material objects then we can save the second law of thermodynamics. And that is very useful, since the resulting book-keeping allows us to work out the efficiency of various energy generation processes involving black holes and ordinary matter.

We can also see that the gravitational energy of the black hole may be considered as residing in its surface. The force holding together a rain drop, its surface tension, results in a surface energy of the same type. If the drop is given further energy it may oscillate so strongly as to break up, due to its being unable to increase its area enough to accommodate the added energy. If further energy is added to a black hole similar oscillations may occur for a similar reason, with even a loss of energy by emission of gravitational radiation. The final state cannot be that of a number of black holes, but only of a single one of area at least that of the original event horizon. Thus the surface energy can never be reduced; we may call it the irreducible energy of the black hole. This is a good name for what we earlier identified with the area of the black hole, since we will never be able to reduce it. In other words, once a black hole of a given area has been formed the irreducible energy stored in the surface stays there till the end of time.

Let us extend the notion of black-hole energy to the case of spinning and electrically charged black holes. We can expect that these additional effects will contribute further energy to the total swallowed up by the collapsar. For example, the black hole would have had to perform work to collect up all the electrically charged particles which it absorbed. This is because they are repelling each other mutually by their electric fields; they resist being brought closer together. The process of condensation of these charges would thus store energy in their associated electric field. We might hope to be able to extract this energy at a later date through annihilation of the charge on the black hole by letting it absorb an equal but opposite amount of charge. Certainly we can identify a term in the total energy of the collapsar which corresponds to this electrical contribution; it has a similar form to that describing the energy on an electrically charged metal ball.

We can similarly recognize a rotational energy contribution which would have the same nature as that in the rotation of a mill-wheel or roundabout, its size depending crucially on the speed of rotation. In all, then, we can see that the energy of a black hole is the sum of the irreducible gravitational energy residing in the surface of its event horizon, the rotational energy depending on its rotational speed, and the electrical energy generated by compressing the charges it contains inside its event horizon. The rotational and electrical energies are not expected to contribute to the entropy of the objects, since they can be increased or decreased at will, by slowing the black hole down or neutralizing its electrical charge. Thus the entropy arises purely from the surface energy.

Now that we know how to split the energy of a collapsar into different types and have determined which of these types can be increased or decreased consistently with the second law of thermodynamics, we can turn to determine the maximum amount of energy which can be extracted from one or more black holes. We will consider first of all only two of them, and calculate how much of their total energy can be extracted by their coalescence. As would be expected the result depends on whether they are spinning or not, charged or not.

First we determine the result when both of the black holes are neither spinning nor charged. Then the highest efficiency obtainable is about 30 per cent, when they are both of equal size and their final coalesced state is a neutral non-spinning collapsar. In other words the final object is only a little more than two-thirds as massive as the total mass of the two initial black holes. The remaining energy, one-third of the total mass, will be emitted as gravitational radiation; it is this which is the available energy, and which we will have to trap if this process is used as a source of power.

The efficiency increases if we take two spinning black holes or two charged black holes. In either case if they are either spinning in the opposite directions but with equal rates of rotation or have exactly opposite electrical charge they lose the maximum possible energy if they coalesce to form an electrically neutral non-spinning black hole. The total mass loss in this case is half their total initial mass; it is as if one of them is completely annihilated. The resulting efficiency of this as an energy generator is evidently 50 per cent, a considerable increase on the previous cases.

The increase has been achieved because of the fact that the energy of the initial black holes was composed of as large a percentage of rotational or charged energy as was possible. One might think that provided the collapsar is spun rapidly enough or is electrically charged to a high enough value then an arbitrarily high proportion of the object's energy could be put into the rotational or charged form, respectively. But in doing so we would be forgetting that if that happened we would be violating some very basic laws which might get us into very grave trouble indeed.

If we try to spin the black hole round fast enough a particle residing on its event horizon could be travelling at faster than the speed of light. We know that energy cannot travel faster than light without giving rise at least to causality paradoxes; if a black hole is made to spin so fast then far, far worse things can happen. The event horizon, shading our eyes from the horrors of the black hole interior, would be annihilated, and we would see the ultimate destruction of matter at its centre.

In a similar fashion if we try to push too much electrical charge into a black hole its gravitational attraction will be insufficient to prevent the charge from spreading itself out again; keeping it in the one-way event horizon is also expected to have similar dire consequences.

So if we try to prevent a Pandora's box opening up we are restricted to a 50 per cent efficiency in the energy generation from spinning or electrically charged black holes. We can, however, increase that to about 65 per cent if we take to begin with two black holes, both of which are spinning and electrically charged. If they are chosen to spin in opposite directions but at the same rate, as before, and also to have equal but opposite electrical charge, again as earlier, then we obtain the figure of efficiency just mentioned provided the final black hole which results from their fusion is not spinning or electrically charged.

Actually it is possible to improve this efficiency if there are more than two black holes. Suppose we have four equal collapsars. Then if we fuse them together in pairs in the most efficient ways described earlier we obtain 30 per cent, 50 per cent and 65 per cent of the mass as available radiated energy in the neutral non-spinning, the charged or spinning, and the charged and spinning cases respectively. If we then collapse the pair of resulting objects to get a single black hole in the end this produces further available energy; the net resulting efficiency is now 50 per cent, 75 per cent and nearly 90 per cent. If this is done for, say, eight black holes the efficiency is even further increased to 65 per cent, 88 per cent and 96 per cent respectively. There is, in fact, no limit to the efficiency of energy extraction in this way, provided we have enough black holes at hand. As we mentioned in the previous chapter there may well be a wealth of them in our galaxy, so that fusing a large number of black holes together may be a very efficient source of energy.

This possibility had already been mentioned at the end of the previous chapter as an explanation of the origin of the fantastic energy generation in the quasi-stellar courses, the quasars. It was suggested there, that these stellar-like very distant objects have a central black hole which is emitting various forms of radiant energy by swallowing up surrounding matter. But it becomes an even more efficient energy producer if the quasar was originally a galaxy of heavy stars which became black holes in the course of evolution. There may have been so many then that they began to fuse with each other, and that is the process we may now be seeing. In fact such a model, with electrically neutral, non-spinning black holes, would have a roughly constant rate of energy generation if the fusion of black holes proceeded at a uniform rate; it appears to be a reasonable explanation of quasars.

But then we have to determine if our own galaxy will suffer the same fate, to become a seething mass of coalescing collapsars, gobbling each other up with the emission of a very large amount of radiation. We obviously cannot say at present with any accuracy if that would happen, but even if we have only one black hole in the galaxy our ultimate fate is a gloomy one: we will be swallowed up by it in the end, as we described earlier.

The energy generated by black-hole fusion could be very difficult to collect, especially if it were all in the form of gravitational radiation. It would certainly be hard to catch the energy released from a chance fusion of black holes; we would need to be able to control the place and time of coalescence so that we could surround it by enough radiation collectors that a good proportion of the released energy were caught. We would also have to be careful to avoid being swallowed up by the black holes as they move together. This would be easier the smaller the black hole, since the total surface area needing to be surrounded would be kept to manageable proportions. But there would be another difficulty arising then.

Black holes are expected to form naturally from an aggregation of matter which is heavier than about twice the sun's mass. The heavier the initial object the sooner it will burn up any nuclear fuel in its interior and collapse to form a black hole. But if a smaller amount of material is taken then it will not collapse naturally; its internal combustion will be sufficiently strong to resist implosion due to the gravitational attraction between its parts. To make a black hole of a smaller amount of material would require very strong external forces to drive its components towards each other to a sufficient extent that all of the material would be compressed inside its event horizon. It would then have become a black hole.

In order to make a small black hole we would need to implode the initial material by an enormous amount. Suppose we wished to compress sixteen hundred tons of iron inside its event horizon. Then we would have to reduce its size to one hundred billionth of a centimetre, a feat requiring a very large expenditure of energy. A calculation shows that to achieve this would require one hundred million times more energy than could be extracted from the iron, even if there were 100 per cent conversion of its mass to energy (so not making a black hole at all). This enormous amount of energy could only be obtained by nuclear fusion processes which use all the deuterium in the oceans of the world. Even then it would be difficult to prevent a proportion of this power being used for compressing the iron from escaping as heat or radiation before the black-hole stage is reached. At first glance then, making small black holes does not appear to be a very useful thing to do.

Before we can dismiss the black hole as a usable source of power in the foreseeable future, we need to consider if there are other methods which will allow us to extract energy from them. Indeed there are, one of these having some technological feasibility. It is a process which essentially extracts the rotational energy in the case of a spinning black hole. If we suppose it were possible to build some form of a stationary structure round a black hole then lowering a particle on a spring deeply enough will allow us to extract extra energy from the system in the form of oscillations of the spring. To achieve this the particle needs to be suspended in the ergosphere, the region between the frozen surface of the black hole as seen from far away and the one-way event horizon further inside it. Here we have a way of extracting the black hole's energy of rotation, so slowing it down in the process. We do not really have a perpetuum mobile, since we get no more useful work once the black hole has stopped rotating. The efficiency of this process may not be very high, however, since the spring would have to be very strong to withstand the forces on it near the black hole.

It might be more practical to extract this energy by a ballistic method. This involves firing a projectile into the ergosphere so that once inside, the projectile explodes into two parts, one of which is ejected out of the ergosphere, the other shot inside the event horizon to be lost for ever. If this is done in the correct manner then the ejected particle can be more energetic than the initial projectile; energy will thus be gained.

The amount of energy we can extract in this way naturally depends on the proportion of reversible energy stored in the black hole, in this case in rotational form. The most extreme case is when all the energy is in this form, and the area of the one-way event horizon is then zero.

We might think that we could extract all of the energy from

such a collapsar, leaving it no longer collapsed, but that would be difficult since the projectiles falling into it will cause it to acquire an irreducible gravitational energy. In any case it could be difficult to make a highly spinning black hole, so that the amount of available energy would be expected to be much lower than the total present in the system. In a similar fashion we expect to be able to extract the electrical energy from a charged black hole. One way to do this would be to use a charged projectile which splits, like the one for extracting rotational energy. It would have to have opposite electrical charge to that of the black hole, so that the projectile is attracted by it. Once in the ergosphere it would split into an ejected neutral part and the charged portion be sacrificed to the black hole. As in the rotational case there is a limit to the amount of energy which could practically be extracted from a charged black hole, particularly since a highly charged one is difficult to make.

It is apparent from what has been said that black holes can only give up their rotational, or electrical, energy and no more. They are not inexhaustible sources of energy. But then we don't expect anything to be so; you cannot get something for nothing in this Universe. The best way to use black holes as energy sources would thus be to take the energy from already existing ones. We saw earlier that it is very hard to make small black holes, and in any case they won't then be able to provide much energy.

We can envisage a technologically very advanced intelligent civilization which goes in for black-hole farming. To do this they spread hydrogen or helium throughout a region of a galaxy, or concentrate some already there, so that large stars are formed rapidly. These are then used as energy generators during their nuclear burning phase, and allowed to collapse to black holes, also connecting supernova energy emitted during the short implosion time. The resulting black holes are then brought together in pairs by suitable methods to obtain a large fraction of their available energy. The resulting single black hole is then finally exhausted of all its remaining available rotational and electrical energy.

The amount of energy available in this way would be

enormous, the most rewarding stage in the farming – the 'harvesting' period – being the time of the black-hole fusion. Of course there are side-problems, especially that of pollution: how to dispose of the single black hole remaining at the final stage. It would have to be stored in a special place, away from the centres of civilization, and would undoubtedly grow so large that the galaxy would have to be evacuated before it was completely swallowed up by it. Who knows if the quasars are not now suffering that very fate, having been born long ago and subsequently squandered by their intelligent inhabitants in this way?

It is possible to convert some of normal matter into energy by using a black hole. To achieve this a chunk of matter is thrown into the black hole so that it spirals down to the event horizon, ultimately being captured by it. As it falls closer it emits gravitational radiation; before it has disappeared for ever inside the one-way membrane, it has radiated just over 5 per cent of its mass. This gives a five-fold higher efficiency than by using nuclear fusion processes, and as such is very worthwhile considering seriously. It might even be possible to increase this efficiency to 40 per cent by shooting particles at the black hole so that they are only just trapped, and spiral into the event horizon at the slowest rate possible. The energy would appear as gravitational radiation, so again one has the difficulty of its collection. But its advantage over previous methods is that it is only limited by the resources of normal matter. Used in this way a black hole is like a hot fire; all it needs to keep giving out energy is more fuel.

There are various further ways of extracting energy from black holes, and even of making a black-hole 'bomb'. They basically depend on the fact that light and other waves can so scatter off a black hole that the impinging wave is amplified, at the expense of the rotational energy of the black hole. This 'super-radiant' scattering has a very low efficiency, the maximum being a few tenths of a per cent for a black hole rotating at its maximal rate.

The process can be turned into an explosive one if the black hole is surrounded by mirrors, so that each amplified scattering is reflected back into the black hole and the amplification repeated. The energy stored in the radiation trapped by the mirrors would then grow extremely rapidly, at the same time draining away the collapsar's rotational energy. Nature may even provide her own mirror, the radiation being reflected off a plasma of interstellar particles trapped around the black hole but prevented from falling into it by the radiation pressure. What vast explosions are occurring through the cosmos when such powers are unleashed?

Because of the difficulty in making them, artificially produced black holes would not appear to be of value even to the most advanced civilization and certainly not to ourselves. The most important type of black hole for the future is the naturally occurring one. If there are as many in the heavens as has been suggested, say over half of matter being in such a collapsed state, then at least we can get into the harvesting period without having to grow our own black holes.

We have to be careful of blanket statements about what can or cannot be done in these matters, however. Only forty years before the first atomic bomb there was no inkling that such power was available, and even ten years before its construction a renowned physicist said it was impossible to make. If we allow for the possibility of artificial black holes being made on earth we have to recognize that they contain the threat of our own destruction.

A black hole weighing sixteen hundred tons, if let loose here, would immediately sink to the centre of the earth and promptly devour it, and us included. To prevent this the artificial collapsar would have to be trapped in some way. It would seem impossible to control it by gravitational forces, unless it was constructed in space, away from the earth's influence. Since this latter way of producing it would add enormously to the production problems it would seem better to consider a way to control it once it was created here on earth.

The only useful 'handles' a black hole has are its mass, spin and charge. It only seems possible to use the latter to prevent it sinking to the earth's centre and so destroying us. If it was made with the maximum possible charge it could be held by an electric field strength of one hundred thousand volts per centimetre. It is remarkable that this field strength is independent of the mass of the black hole, but this is due to the fact that as the black hole gets heavier so the electric charge it can sustain increases; this increases the response of the charged black hole to an electric field. A field of such strength can be produced and even maintained. One of the most severe difficulties in keeping high-voltage supplies arises in the breakdown of the insulating material around high-tension cables. This has now been avoided by the manufacture of suitable materials which have been utilized for power transmission at over one hundred thousand volts. Since it turns out that the higher the voltage the more economical the energy transfer process, there are even now materials being developed to allow transmission at one million volts; the insulators will break down and pass an electric discharge at an electric force of about one million volts per centimetre.

This is encouraging, though a minute black hole in the middle of an insulator will start to gobble up the material, producing heat and radiant energy in the process. It may well cause the insulator to break down too rapidly, though one would expect it to isolate itself in the centre of a sphere of energy and so drive other material away. However, the best material to use would be none at all, in other words a vacuum, since then the problem of electric breakdown would be almost absent, as would the chance of the black hole absorbing extra material (and thus becoming heavier).

A saucer-type of electric field profile would achieve the necessary stability. However, many safety precautions would be necessary to prevent failure of the containment mechanism, even going as far as having the whole system mounted in a space-craft distant from the earth, so that failure would not be catastrophic. Such may be the basis of the ultimate rocket drive motor.

The other reason we must try to see how far we can stretch our minds by considering future possibilities is that it is good for us. To have one's mind boggled at least once per day is a very important part of increasing flexibility and dispelling mystery in our attitude to the world around us. But the mystery of the black hole seen from outside would seem to be absolute. We can never expect to find out what goes on inside it. Even if we extract as much energy as we can from it we still might consider it as the ultimate, the infinite. So to enlarge our view of the possible we now come to discuss the impossible: what it is like inside the black hole.

7 Inside the Unknown

The event horizon of a black hole is truly the boundary of the unknown. Outside it is the normal world in which we live and on which we can experiment. But once inside the horizon it is never possible to communicate with the outside. Even if some experimental scientists were brave enough to venture inside the black hole, we outside would never know what they had found. So how could we ever discuss what would happen to them? Well, the answer is that we cannot do so with absolute certainty. But in any case much of what has been said so far about collapsars has only been based on theories which we know to be good in the vicinity of our solar system. We have not been able to use direct experience of black holes. In a similar way we can build up a picture of what we would expect to find if we did go inside the event horizon of a black hole by applying our theories to that situation. We cannot be certain that our ideas will still be valid there, but at least they will give an initial indication of what to expect.

The first thing to realize is that as we dive through the event horizon the frozen star would suddenly vanish as we were about to strike it. This is due to the light it had emitted as it collapsed through its own horizon still leaking out and making it look as if it were still there. Of course the frozen image of the star we would seem to be about to hit would be very dim, but it could well cause apprehension. That would be the only strange occurrence as we fell through the one-way membrane, at least for a large black hole for which the gravitational forces on us were still not excessively large at the horizon, say only as strong as those on the surface of the earth.

In that situation we might not even realize we had actually fallen through the one-way membrane – nothing would go 'bump'. But yet once inside the horizon we would find a topsy-turvy world indeed, one of the strangest we could ever envisage. For in it time and space would have been interchanged. In our normal world we can move freely about in space, provided we have the energy. There is no limitation on our motion in any direction we choose. Time is just the opposite; it rolls on inexorably, and even though we might slow it down near the surface of the event horizon it still marches onwards.

This is exactly reversed inside the event horizon. There we would have no control over our voyage through space, even though we could go backwards in time, or simply move around in it to our heart's desire. But our space-ship is drawn inexorably downwards towards the centre of the black hole, and no power we could conceive of can prevent us from being pulled ever deeper into the centre of the black hole. The inevitable passage of time outside the event horizon has become the inescapable voyage to the centre of the black hole.

It might be hoped that if time has been freed from its shackles inside the event horizon then we might be able to make our voyage to the black hole's centre last as long as we like. But each of us carries with us his or her own personal time, called our proper time. It is that which would be measured by an accurate watch or other timepiece which was moved around with us. My proper time could be very different from other people's, depending on various factors. It certainly is altered by being near the event horizon of a black hole, at least as seen from afar; a single beat of my proper time in that situation would be equal to billions of beats of a distant observer's proper time.

We have already discussed the manner in which time depends on how fast you move, since 'moving clocks go slow'. The equivalence between the strong gravitational attraction experienced near a black hole and an acceleration away from the event horizon led us to understand why a clock near the surface of the collapsar is retarded in comparison to one very far away.

We can also see why this slowing down of time occurs near a black hole if we consider what happens to light as it tries to escape from just above the event horizon. Light possesses energy, and so has mass. It is therefore acted on by a gravitational field, as can be seen, for example, when a ray of light is bent towards the sun as it flashes past it. In order to escape from near a black hole, then, light will have to work very hard to overcome the force of attraction exerted on it pulling it ever down into the mysterious event horizon.

When the radiation does finally escape from the vicinity of the black hole it will therefore be found to have a much lower energy than when it was just emitted. It is known that the energy of a beam of light is proportional to its frequency. So light which has finally reached safety after being radiated close to an event horizon will have had its frequency reduced. If each pulse of the light is used to trigger a clock mechanism then the rate of passage of this time will be slower than that ticked off by a similar clock using light from a nearby source.

Light can never escape from inside the event horizon; so it might be difficult to relate the times inside it to those outside. Even so it is possible to measure the total proper time it would take to fall into the very centre of the black hole from its event horizon. This time turns out to be finite, naturally depending on the size of the black hole; the larger the hole the longer the time. So if we were falling into a black hole we would be driven into its centre after a certain amount of our time had elapsed, however much we squiggled and squirmed and fired our rockets to escape being crushed out of existence into a point at the centre of the black hole.

It is reasonable to ask what it is that impels us to the collapsar's centre, with absolutely no chance of escape. Why is there no force strong enough to enable us to escape? Why is there no chance of a collapsing star avoiding ever further collapse till it is squeezed to a single point? In fact the answer is that any such force trying to prevent further collapse must contain energy to exert an effect. This source of energy will itself act as having a mass, since energy and mass are equivalent, '$E=mc^2$'. But then there will be a further gravitational field exerted by this mass, which will even hasten collapse. The larger the energy trying to prevent collapse, the greater the gravitational field producing further compression. It is just not possible to escape this effect. Collapse to a point will always result, however frantically one tries to avoid it.

Once inside the black hole, then, the fate of complete annihilation can never be avoided, nor can it be delayed beyond a certain length of time. For a small black hole, say just a little heavier than twice the sun's mass, the time to fall to the centre is exceedingly short, being about twenty millionths of a second. But we would be dead by then in any case, torn to pieces by the strong gravitational forces already present at the event horizon.

It is only when a black hole of much larger size is investigated that any noticeable time elapses between falling past its horizon and being destroyed at its centre. In a black hole a million times as heavy as the sun it would still only take ten seconds, while in one a billion times as heavy as the sun it would take three hours. Only in one as heavy as our galaxy would we begin to have any appreciable time, but it would still be only about two weeks.

We have to remember that the time we experience wherever we are, be it here on earth or falling to our death inside a black hole, is our proper time, the time measured by a wristwatch we carry with us. It is necessary to emphasize this, since it is not correct just to talk about time without any qualification as to how it is measured. This extra specification is doubly necessary inside the black hole, where the traditional notion of space and of time are interchanged.

We may attempt to use the same notion of time as that in operation far outside the event horizon. We then find that as we fall towards the one-way membrane that time increases without limit. If we could observe the distant clocks far away from the black hole they would appear to go ever faster as we approached the event horizon, till ultimately we would not be able to see their hands moving over the clock faces. As we pass through the event horizon the hands would have to reverse and start slowing down as we approach the centre of the black hole. We are apparently going backward in time!

Such a bizarre notion of time is evidently not relevant to experience inside a black hole. It can be used as a possible label for indicating roughly 'when' we experience various effects. But if we want to use a time which goes forward at a decent rate, that of our own bodies, then we need to turn to proper time, as we said earlier. It is that, especially, which will tell us how close we are to our fate at the centre of the black hole.

As we fell to the centre we would be able to notice a very interesting phenomenon. We could send messages to space-ships falling ahead of us but they would not be able to reply to us. Nor would we be able to respond to communication from any space-ship falling later than ourselves. It would be as if we were passing through a series of one-way membranes surrounding the black hole's centre and getting ever closer to it. That is an effect we would have to be prepared for if we wished to travel to the centre in company with other space rockets. In order to be able to communicate at all with each other we would have to stick as close together as possible and try to stay exactly the same distance from the centre. But, as we see, that would be easier said than done.

The actual nature of the world inside the event horizon can be even stranger than this. The enormous distortion of space could become so extreme as to generate a 'throat', which may be connected to another universe. Not only time but also space is very different there from our normal experience of it!

This throat, or bridge, to another world is not expected to occur for a normal collapsing star, but it could have been generated at some time in the past. It arises as a special solution to the equations of Einstein's theory of general relativity and was first obtained by Schwarschild in 1917, a few months after Einstein formulated his theory. This Schwarschild solution is of special interest since it contained the first description of a black hole.

The curvature of space in the Schwarschild solution can be described very accurately. It can be visualized best for flatlanders, a land inhabited by people so thin that they have height and width but no thickness. The world would be truly flat, very distant from the centre of the Schwarschild black hole, but as the flatlanders approached nearer to it it would be as if the land was sliding down the edge of a funnel, getting smaller and smaller, and more and more curved. The double funnel obtained by joining another funnel to the first at their ends now completes the picture. A flatlander's journey to the centre of his universe would continue with his passing through the most curved part, the 'throat', of this double funnel, and so out into an ever flatter landscape just like the one he was originally in.

Ultimately it would be flat again far away from the throat and he could settle back in peace. But he would now be in a totally different flat world from the one in which he started.

Here is a puzzle indeed, the duplication of a single universe by the power exerted by gravity. This effect has so disturbed some scientists that recently they have claimed that these two worlds must ultimately join up when they are very nearly flat, far from the centre. But we do not know if that further bridge must occur; such twin universes have never been seen. Nor do we expect it to be noticed very easily, since to cross from one world to the other across the throat without being crushed to death by intense gravitational fields at the centre can only be achieved by travelling faster than light. The superiority of the speed of light to all other material things is one thing which is sacrosanct even inside the black hole, so we may never know about the other universe joined so phathologically to ours, if it is described, as it could be, by the Schwarschild solution.

While we may be interested in the curious features of the black hole's topology we would undoubtedly be most con-concerned about our fate as we approach the centre. As we do so the gravitational forces on us would be ever-increasing, being ultimately infinite at the centre. We can only withstand a certain amount of gravitational force on us and then we would be killed; we must prepare ourselves for death within a short time.

Knowing the inescapability of this death we might decide to worry about it no more, but use what little time that was left (if any) to compose ourselves to meet our end. Given human nature at least some of us would be more likely to spend our time trying to find out exactly when and how we would die.

The forces on an astronaut's body which would ultimately destroy him, are tidal in nature. Thus there would be a strong attraction on the man's feet and, if he were falling feet first, a lesser attraction on the parts of his body which are further away from the centre. As he drops further and further down these forces increase indefinitely, and so does their difference. It is this latter which will be most operative in killing him, since it will stretch his body out to have indefinite length. However at the same time his volume will constantly be reduced as he falls due to the general compression which is going on at the centre of the hole.

It is possible to calculate roughly how far away from the black star the astronaut will succumb to the mounting tidal forces and the bones and muscles of his body be elongated beyond breaking point along his length and be crushed beyond breaking point in all other directions. For a space-man of average weight and with bones and muscles of normal strength he need only be one hundred kilometres away from a black hole of one solar mass in size before he is killed; he would be still a long way from its event horizon, which is only one and one-half kilometres. He can just get inside the event horizon of a black hole one thousand times heavier before he is killed by it, while for one a million times heavier he can get one hundred times closer to the centre than the event horizon before he succumbs. For a black hole of the size of our galaxy he can get about a million times nearer the centre than the critical radius before his end; in these last two cases the forces on him at the event horizon itself are negligible.

From these values we see that a black hole no heavier than ten to one hundred times the size of our sun can already exert an uncomfortably high tidal force on any space traveller venturing near to it, but still staying outside its event horizon. Such black holes will not present such a threat to the space-voyager of the future as long as his propulsion unit doesn't cease to work when he is passing close to one and so ultimately be captured by it. It will be the large black holes which are really perilous since for them the danger signals of strong tidal forces do not arise till a space traveller is well inside their event horizon. But if these black holes are anything at all like quasars, emitting large amounts of energy from objects they are ingesting, then they would also be very noticeable. We don't know that they will be, however, so future space travellers must still be very wary of unexpected black holes.

The experiences of an astronaut as he falls to the centre of a black hole can be even stranger in the case of a rotating black hole. Decidedly new things can happen to him which make the previous case seem normal by comparison. He will only actually be destroyed by infinite tidal forces in the black hole if he falls into it from an equatorial direction (taking the axis of rotation as north-south). Even that cannot occur if there is also an electrical charge on the black hole, and then only light, falling in the equatorial plane, can ever experience these infinite tidal forces, usually called the singularity of the black hole.

The singularity in a non-rotating electrically neutral black hole is at its centre, and is ultimately experienced by any unfortunate who has erred inside the event horizon. But a rotating black hole (with or without electrical charge) has its singularity in the form of a ring in the equatorial plane. It is this ring singularity which gives the inside of the black hole further remarkable properties.

The fate of an astronaut falling equatorially into an electrically neutral rotating black hole is clear: he will ultimately be killed by having to approach arbitrarily close to the ring singularity in a finite amount of his own experienced proper time. But any other entry to the black hole, or any entry into a charged rotating one, need not end in the astronaut's death unless there are already high tidal forces near the event horizon. That is only the case for black holes not much heavier than about one thousand solar masses. We would expect the majority of collapsing stars to be in this category, so for the average rotating black hole a captured space-man will already have been killed before he even gets to the event horizon, as was so for the non-rotating case.

Let us turn to the much larger black hole, say approaching galactic size. A falling astronaut could then remain alive throughout his voyage into the event horizon. Yet he cannot ever return to the outer world again. What happens to him? Does he continue to go round and round inside the one-way membrane till he is either starved to death or dies of old age? The answer is the most surprising thing of all. In the earlier part of his journey through the event horizon he has experienced the same switch of space and time which he would have met throughout the non-spinning neutral black hole. But in the rotating case after he has been driven inexorably down towards the ring singularity he falls into a new region in which space and time are interchanged again, so that time now flows inexorably on, and space is under his control again.

At this stage he might breathe a hearty sigh of relief. But his troubles have only just begun. The region he has now entered is in fact joined on to another universe altogether. He certainly can now begin to move away from the singularity, and should be able to communicate with his companions falling with him. He may even try to ascend through the topsy-turvy region he had previously fallen through. But if he did try, he would in fact have to travel through a region with similar properties but in a different universe. Try as he might he could never return to his old universe, even if it were the topsy-turvy part of it inside the black hole.

The trouble is that once inside the rotating black hole he is driven down through the topsy-turvy region and expelled from that into a no-man's-land, with usual space and time properties. It is this no-man's-land which contains the ring singularity he avoids, but it is joined up to two universes, his and another. And once he has left his own he can never return to it, but must go on to the other. What he will find there we have no idea in general.

There is one thing it will contain, however, and that is the identical rotating black hole. The astronaut might long to return to his own universe, and think that falling back into the black hole would enable him to achieve his desire. He would soon be sadly disillusioned, however, for he would pass through a copy of the topsy-turvy land into a different no-man's-land and thence into a further universe, different from the first two. Yet it again would certainly have the identical rotating black hole in it. And so he could continue his journeys going from one universe to another but never returning to his original one.

This remarkable feature is lost when non-spinning black holes are involved. They have the decided advantage of not turning a space-traveller into the wanderer, ever parted from his loved one; they unfortunately compress all they receive into nothingness, mere points with no existence and certainly no flesh or blood. Evidently the latter type of existence is less attractive than the former, and it is lucky that we expect the average black hole to be spinning.

There are further surprises in store for the space-traveller who has fallen into a rotating black hole. Once he has fallen out of his own universe into the no-man's-land at the collapsar's centre he has the possibility of going on a time journey. If he moves in a circular path around the line about which the black hole is spinning, but in an opposite direction to that of the spin, he can make up as much lost time as he likes. On each circular trip round the rotation axis the astronaut gains an amount of time proportional to the spin of the black hole. Of course he cannot effectively use this time he has gained, since he cannot return to his own universe but must go on to another. Such a possibility of time travel can only be recommended to a fanatic; for the average person, he should keep well away from such an experience!

We would appear to have stumbled on a paradox, since time travel would seem to imply that we can go faster than light. At least the converse of that idea is true: a faster than light traveller can return to his or her place of departure at an earlier time than he left. In the words of the anonymous poet:

> There was a young girl called Bright
> Who could travel much faster than light
> She went out one day
> In an Einsteinian way
> And returned the previous night.

This effect happens because Miss Bright could travel faster than any information, and so could effectively 'beat the clock'. But the converse of this is not true: time travel is possible even if it is performed at a speed always slower than light. The effect arises because of the curious nature of the space inside a rotating black hole. It behaves as does ordinary space if two places in it were identified; standing at one of these points immediately transports a person to the other point. This identification of certain parts of a space with other portions of it can be achieved legitimately, without destroying the never-faster-than-light condition. But it certainly produces some bizarre effects, as we have seen.

As with the twin universes mentioned earlier in connection with the Schwarschild black hole, we do not know if it is possible ever to penetrate deep inside a rotating black hole. There have recently been suggestions that after travelling through the topsy-turvy land inside the event horizon it would not be possible after all to fall right out of it and into what would appear to be a normal world, with space and time being the usual way round, though still possessing the bizarre possibility of time travel. Calculations have indicated that a falling space traveller would be crushed to bits just as he fell out of the topsy-turvy land back into the sane world. He is expected to suffer just as at the very centre of the black hole, being crushed and torn to death by infinitely strong tidal forces.

The rotating black hole may not be as gentle as was suggested. But here again we will never be certain of the answer, since whoever tries to find out by falling into it will never be able to tell us. It is just like death and immortality – nobody has ever come back from the grave to tell us convincingly enough what it was like on the other side. And if they ever came out of the event horizon round a black hole then we really would have to believe in ghosts!

The very fact that such journeys in time are possible is very disturbing. It could destroy the reasonableness of the universe that we inhabit. We do not, in fact, ever meet anyone identical to ourselves who has travelled back in time from some distant future when the time machine had been invented. This is one reason why the development of such machines in the future would appear impossible, at least in our world. Of course the particular no-man's-land between universes in which such time travel is now a possibility for a dare-devil astronaut is not connected back to our own universe. So such a disconcerting meeting outside the black hole is not possible.

Inside the no-man's-land, however, the explorer could well be startled out of his wits by suddenly seeing a copy of himself coming towards himself as he passes through the surface enclosing the no-man's-land around the ring singularity. He could even meet a whole host of himselves. But then which one is his true self? Are they all equally legitimate, or is the first the best? And if so what about the others?

These questions are very perplexing, but at present there appear to be no answers to them. Our whole way of thinking has been conditioned to exclude such a possibility. Nor do we need to change our attitudes if we only wish to understand our present 'normal' world. But the spinning black hole very likely exists, so that somewhere, somehow, it would be possible to go through such strange experiences. We should face up to the probable, even though it appears to give rise to the impossible.

8 A New Start?

The impossible becomes even more manifest at the very centre of the black hole, at its singularity, so to say, where the gravitational tidal forces become infinite, either in the spinning or stationary case. For there objects – astronauts, spaceships, and anything else, are crushed out of existence completely. An astronaut, falling in his ship to the singularity at the centre of a stationary black hole, would be killed before he reached it. But the pulverized mess that was his remains would be crushed ever smaller and smaller till it reaches absolutely no size at all. And this takes only a finite amount of time to occur, as measured by a very strong timepiece embedded in the mess.

The problem is then: what happens after the singularity has been reached? The laws of physics which we have been using so far to describe the black hole are no longer sufficient; they cease to predict how the ever-more compressed object will behave after it has been reduced to zero volume. Of course if the majority of black holes are spinning then only very few disasters of this type will happen, being exactly when the astronaut falls in an equatorial plane into the ring singularity. So such a difficulty may not be met for a very long time, if ever.

The problem we are faced with is still a serious one in our attempt to understand the impossible made manifest by the black hole. We have already met several very bizarre situations, but this is the worst of all. For the laws by which we think that events in the world around us are ruled are no longer operable at the black-hole singularity. New rules must be discovered which will allow us to continue describing and understanding the developing universe.

It has been claimed by some scientists recently that this

difficulty raised by the black-hole singularity is the most important one for science in the whole of the twentieth century. A similar type of situation arose at the end of the last century, when calculations based essentially on the laws Newton had proposed two centuries earlier gave infinite results for the amount of energy radiated from a hot oven with a hole in it. To remove such an absurdity, Planck, in 1901, introduced discreteness or quanta into the hitherto continuous world of Newton. In so doing he opened the door to the complete replacement of Newton's rules of mechanics by what is naturally called quantum mechanics, in which the certain world we see around us is replaced by one in which nothing is certain, nothing is as it seems. We now know that all matter is composed of waves of probability, and the certitude of the classical world has vanished.

This was total destruction of a view of the world which had persisted since the time of the ancient Greeks, when man began seriously to study and think about the nature of existence and of the world in which he struggled to find meaning. This change has had enormous repercussions in many ways, not least producing technological developments of which the atomic bomb was one.

We are now faced with another paradox. It is possible that quantum mechanics will save us again. The ideas of uncertainty would possibly help by smearing out the point-like precision of the black-hole singularity. Nobody knows yet if it will do so, though there are indications that it may only make the situation even more singular. But certainly it is essential to take account of these uncertainty effects to discuss the final stages matter passes through as it collapses to zero volume at a black-hole singularity.

One of the ways of describing how such probability features of the world may be very important is by means of superspace, introduced by Professor John Wheeler of Princeton University, who can be regarded as one of the 'fathers' of the atomic and hydrogen bombs. The certain motion of a football from one end of the field to the other in Newton's mechanics has to be replaced by a whole host of its possible trajectories in quantum mechanics. Some of them will be very unlikely, such as travelling first to the sun and then back to the other end of the field; others will be much more probable if they are very close to the Newtonian motion. That is why the Newtonian prediction about where the ball goes is as good as it is, and satisfied men for nearly three centuries.

Superspace is the similar plethora of possible developments of the space in which we live. Some of them will be very unlikely, with hideous distortions which would send a normal human mad if they were experienced. Other developments are so strange that it would be almost impossible for us even to recognize them, they being so far from our previous experience. As in the case of the football the most likely development of our surroundings under the influences of the gravitational forces between its parts is the one described by the equations of Einstein.

Near the singularity of a black hole quantum effects will become important and the plenitude of worlds in superspace should begin to be experienced. But the superspace being explored ever increasingly as the singularity is reached has none of the properties we would expect. The usual ideas of before and after, of what happens 'next', have to be abandoned; even time itself loses any sense, and we must be prepared for the instantaneous jump from one point of space to another if we can once understand how to penetrate and escape from superspace.

The gates to superspace are not expected to be very large. It may not be till the atoms of the astronaut and his spaceship have been compressed about a billion billion billion times at the black-hole singularity that these gates open and allow the matter to fly freely through superspace. By then even the atoms themselves, let alone the astronaut and his space-ship, will have completely lost their identity. So the possibility of using superspace to achieve instantaneous journeys or even time travel seems rather remote.

The voyage of an astronaut and his ship could perhaps be achieved through superspace if the problem of reconstructing their configuration of atoms from such a highly compressed state could be solved. But this would have to be performed by agents external to the now jumbled matter. These architects would have to be armed with the blueprint necessary to reconstruct the astronaut and his ship. But the plane itself would have to travel at a speed slower than light, so that would be the limit to the speed which the astronaut (though not his atoms) could travel. The atoms might even be able to be sent into the past by travel through superspace, but they would have to wait till the future of the time jump before the blueprint could be obtained which would allow the astronaut to be re-made.

We can get a better picture of what superspace may look like by returning to our earlier model of the Schwarschild black hole, composed of two worlds joined by a bridge or throat. If we join up the distant parts we get a space with a worm hole in it, being in the shape of a distorted doughnut (which turns out to be a good model for a stable cloud or packet of energy). This aggregation has mass, and we can regard it as the fundamental unit of matter in the universe, out of which all we see can be constructed. Such a bundle of energy is called a geon; it has a mass of about a hundred-thousandth of a gram.

We can now picture the world in close-up as composed of many of these geons. In accord with the probability notions of quantum mechanics they are constantly appearing and disappearing under our very eyes. As Wheeler has so eloquently put it, 'Space is like an ocean which looks flat to an aviator who flies above it, but which is a tossing turmoil to the hapless butterfly who falls upon it. Regarded more and more closely it shows more and more agitation, until . . . the entire structure is permeated everywhere with worm holes. Geometrodynamic law' (the law of geometry and dynamics combined) 'forces on all space this foam-like character'.

This partially explains why the traveller into superspace has to leave all his usual notions of space and time behind him. He cannot ask if superspace is hot or cold, whether it is wide or narrow, or whether it is shaped like a cube or a sphere. It has no past or future, nor any dimensions. It is a lace-work of worm holes, forming and disappearing, constantly in motion but never advancing or retreating. It is full of ceaseless activity, yet overall it is static and timeless.

The reason that instantaneous jumps can be achieved in superspace is due to this 'worm-hole' character. If we were flatlanders who lived on the surface of a doughnut it would be much quicker to go from one place on the outside of the ring to its corresponding point on the inner surface by jumping out of the doughnut's surface to get there. Of course we are only supposed to be able to sidle around in a very flat fashion on the ring's surface, so we would be breaking the rules by such a jump.

It is precisely such a jump, through the 'superspace' of the interior of the doughnut, that is allowed by worm holes. They allow these direct connections which would otherwise take so long. We can imagine each of these worm holes as dots giving access to the doughnut's interior at the two points we are interested in, one we can call the earth, the other a distant star. We enter superspace through the worm hole at the earth and leave it at the other worm hole at the distant star; the whole journey could be instantaneous.

This feature of timelessness of superspace is difficult for the Western mind to comprehend, though much easier in the East. There life is to many only one part of an endless chain of lives and deaths, each life followed by reincarnation into a further one. As a Minister of Education in an Eastern country is said to have replied in a complacent fashion to complaints about the deaths of children from smallpox, 'After all, they will be reincarnated anyway'. This view is so foreign to the Western mind that it will find superspace an alien, frightening place, far worse than the lonely Universe in which we live. Our world is hostile, true, but it has time and space and hot and cold, properties we all respond to. All that superspace has to offer is boundless activity in a never-changing world.

There are other possible ways out of the terrible dilemma at the centre of a black hole. It may be that larger worm holes reach out to take the pulverized matter from the singularity and let it bubble up somewhere else in our universe. It is as if there were a tube from the central point that leads out of the event horizon. That could not occur in any simple fashion; we certainly could not expect to see such a tube poking out of the one-way membrane at any point.

The worm-hole type of joining up space in one region to that in another would avoid this difficulty. Such a fount of energy would ultimately be found, so such a possibility is experimentally testable, at least in principle. It may even be that the compressed matter from black holes is reborn by being spewed up inside quasars. There would then be an infinite cycle of birth, death, rebirth . . . and so on throughout the Universe, though only for the material world. Such a picture allows matter to be constantly recirculating, never to be lost. It is the materialist's Nirvana.

We have also to face the possibility that the matter being compressed inside black holes just vanishes, and that is all there is to it. It is almost impossible to conceive of such complete annihilation. We always ask: where does it go to? We have a fixed idea that once created, matter can never be utterly annihilated. The ideas of conservation of some underlying primary substance are strong in us, inherited, no doubt, from the alchemists' dreams of changing the dross earthly representations of this quantity to its attractive glittering form as gold.

Yet if theories of creation of matter can be accepted, why not those involving its destruction? The main difficulty in setting up such ideas is that it is difficult to see how to tie the annihilation process precisely to that involving gravitational collapse. In the theory of continuous creation, suggested three decades ago to explain how the galaxies could be constantly receding from us yet the universe still retain its same composition, matter was to be continuously created to fill up the gaps left by the fleeing galaxies. Hydrogen atoms were supposed to spring into existence from nowhere at a steady rate throughout the Universe. The destructive version of such a theory would have atoms steadily disappearing, just vanishing into nothingness. But there is nothing in such a theory to allow there to be centres of destruction at black holes.

Even if such an idea were possible it would still leave a very difficult problem. For there would have to be annihilation of all the matter in very massive objects; these would produce large-scale effects. Of course, we outsiders would still see the various observational phenomena presented by the event horizon of a black hole – the frozen star covered with its stationary captives, emission of various radiations as objects fall down to this trap, perturbations of the orbits of close-flying space-ships, and so on. But while we would think that there was something behind that one-way membrane, according to the annihilation theory there would not be. Anyone falling into a black hole would still, presumably, fall to its centre to be crushed out of existence in his turn there, but he would not find a jumble of matter in front of him going through whatever would otherwise happen. He would have a clear entry to the nowhere he would go to on his annihilation at the black-hole centre.

This feature of an event horizon still being there, even though there is no longer a collapsed star at its centre, does not seem quite right somehow. We would expect that if the collapsing matter disappeared completely then so should its effects. We may reconcile this with a persisting event horizon by realizing that the distortion of space left in the wake of collapsing matter is so great that whatever happens to the matter at a later time cannot be communicated outside the event horizon. Nor indeed can anything the star experiences at the instant before final annihilation change anything anywhere else; the destruction will be completely hidden from all other regions of space, even inside the event horizon, except at the single point where the collapsar is now concentrated.

Neither of these possibilities of the fate of matter at the centre of a black hole – being spewed out in another part of the Universe or just vanishing completely – has proved very satisfying. The main reason for this is that they both appear rather artificial in the face of attempts to set up complete theoretical descriptions of them. In any case, superspace will have to play its rightful role in the process, and until that has been clarified it is difficult to incorporate either of these further avenues of escape for matter in any more detail.

We don't know what is the resolution to the problem posed by the black-hole singularity. It is a very urgent question whose solution will certainly cause a revolution in science.

But there is an even worse problem that could arise from gravitational collapse; it could even destroy science completely. That is if the naked singularities, unshielded by an event horizon, could be created. Theoretically they are possible, and would arise if the spin of a black hole were so high that its event horizon were moving round faster than the speed of light. The stationary black hole which has more electric charge in it than can be held together by its gravitational attraction is another type of naked singularity.

Such an object as an unscreened singularity would be awesome to behold. There would be now no one-way membrane hiding it from our mortal gaze. No longer would we be shielded from observing the mystery of the final stage of the collapse at the singular centre of the collapsar. It would be as if heaven and hell had suddenly been transported to earth so that we could gaze at them.

In spite of being able to view the terrors of the final destruction of matter, or of seeing it entering the portals of superspace, we might take comfort from the fact that at least we would finally be able to discover what happens inside the black hole at its singularity. We could then throw overboard whatever parts of our science we had to and reconstruct the rest so as to fit what we had seen.

But if a naked singularity could ever be created the whole of science would have to be abandoned. For we would not be able to predict the future, nor even the past. We would be able to travel in time to our heart's desire, to return and meet ourselves, to people our earth with many, many copies of ourself. The time travel which only occurred for the intrepid few who vanished into the no-man's-land inside a spinning black hole would now be available to anyone in the whole of our universe. The trip back in time could only be taken by a suitable voyage round the naked singularity, but it would always be possible.

Scientists have taken alarm at this prospect of their way of life being completely destroyed. They have introduced a 'cosmic censor' who would not allow such chaos to enter their orderly world. Indeed not only they should be worried about the threat a naked singularity poses; we all would have great problems if one existed. Even if it did it might not make any practical difference to us for a very long time. Yet it would and should destroy our faith in the orderliness of our world. We would have to begin to face a universe in which our reason and intellect were not the best tools with which to come to terms with our environment. We have no idea what they could be replaced by.

Preliminary calculations have been made to see if a naked singularity can ever be the end-product of a heavy collapsing star. Fortunately for us they have only succeeded in very special situations which may never occur in the Universe. It is to be hoped they never will. If they do then the orderly universe we thought we were living in is a myth.

The implications of naked singularities are indeed so grave that it is worthwhile to consider the type of situation in which they might be created. We do not expect such terrifying objects to arise from collapse of a star which is not strongly asymmetrical. In the extreme case of the in-fall of a perfectly spherical star then it can be shown, by a very general analysis, that the resulting distortion of space can only be that described by the Schwarschild model of a black hole. Even small deviations from perfect spherical symmetry can be shown to be shone away as gravitational radiation as the star collapses, so that the only possibility for obtaining naked singularities is in the case of the in-fall of a highly asymmetrical star.

There are two situations which have been investigated, and which have a potentially dangerous end-state of collapse. These are first of all for the case of a ball of matter rather in the shape of the earth, with the poles flattened, so being closer together than two points on the equator on opposite ends of a diameter. For this the final collapsed state is in the form of a flattened pancake, all the matter having collapsed onto a plane through the equator. The other case is that of starting from the shape of a rugby football, ending up as a long thin thread.

The story of the collapse of an earth-shaped body to a pancake does not appear to produce a naked singularity, at least as far as the process can be followed. It does reach a stage which becomes very complicated, and while this may develop into a visible singularity it has not been possible to show this so far. The other situation, the in-fall of the rugby football, does indeed appear to give the final state we were looking for (but hoping we would not find!)

The rugby-football-shaped star first becomes more and more elongated, till it takes the form of a long thin thread. The subsequent contraction of the thread's length is negligible compared to that of its thickness. The velocity of this collapse soon approaches the speed of light; the behaviour of the object is very, very different from that which Isaac Newton would have pictured. A detailed analysis shows that this enormously rapid decrease in thickness of the thread produces a singularity bereft of a horizon around it. The thread has thus become a region of infinitely high tidal forces of gravitational origin. The singular thread has now attained an infinite density, and involves an extensive emission of radiant energy in the form of gravitational waves.

The reason why no event horizon is formed around the collapsing thread, nor does it form around the pancake either, is basically due to the fact that the total mass of the star has not been compacted into a small enough region in all directions. This means that information, in the case of the thread, can leak out along the thread, even if it might find it difficult to escape directly away from it. Any event horizon which would try to form would thus always have a hole in it, so defeating the purpose of an event horizon in the first place.

This result means that we should be wary of the collapse of a rugby-football-shaped heavy star. It is that which could destroy our orderly universe. It could be that only by construction of such a monstrosity would we ever learn what really does happen to matter at the final stage of the collapse to a non-existent point. If so, it would be a hard price to pay – the end of our own world – for discovering what such a finale really is. There may be a few scientists who would feel it worthwhile to make the sacrifice, but undoubtedly they would have very little support.

We can reassure ourselves by appealing to our cosmic censor, who never allows such terrible objects as naked singularities ever to exist. But the difficulty is that though such a censor brings security there is no reason for his or her existence; the cosmic censor plays the same role for a scientist working in problems of gravitational collapse that God does to the lost and helpless. Appealing to such principles is not playing by the rules of the game we set up earlier – to look the infinite in the face as long as we can without flinching and turning tail. So we have to dispense with the cosmic censor, and appeal to the universe either here on earth or out in the heavens. There may be a naked singularity out there; one might even be created here on earth if we tried hard enough.

The only comfort we can take is that, other than being foolhardy enough to make our own, we would in all likelihood be living in a part of the universe in which there would still be enough order for us to get by. But that need not always happen in the future. And what if some madman threatens to make a naked singularity unless his demands are met? Truly he would have the doomsday machine, and we ourselves would be thrown into chaos if we did not acquiesce. Could such a fate have overtaken the quasars? All we can do is make a guess, but we realize now that more is possible in this world than was dreamt about in science fiction!

9 In the Beginning

The black hole presents an impenetrable mystery to us, if we stay safely outside it. We can never be sure that we are correct in our conjectures about what is happening inside it unless we are prepared to relinquish our place in this world to fall inside the event horizon, either to suffer certain death or at least to travel to a new life in another world. But what about the world we are already in? At least we should try to satisfy our curiosity about it and clear up all its mystery. We certainly would not seem to be under the same limitations which restrict us when we try to probe the black hole. Yet our own universe presents as great, if not greater, difficulties when attempting to comprehend it.

The hardest question of all to answer is where did our universe come from? If we reply that it came from somewhere else, brought to its present state by the action of the laws of physics, we need only add that somewhere else, filled with whatever was in it before it formed us and our material surroundings, to our present world. We then ask again where did that new totality come from? Any definite answer to our first question is the wrong one, since it would lead to an infinite chain of similar questions. We have already specified that we are talking about everything that is material. There can not be, and never can have been, anything else, out of which our world came.

But we could try to find from what our world, as we know it today, arose. We might do so by conjecturing that our present universe sprang into being from the final stage of collapse in a spinning black hole in a different universe, bubbling out of the black-hole's centre as explained in the previous chapter. That might or might not fit with experimental facts if we looked for them carefully enough, but it would still beg the question, since we have then to explain where the previous universe came from.

The problem is made especially difficult because of the seeming impossibility for us to think of something coming out of nothing. In particular we expect that energy is conserved; if the total energy of the universe is positive then some form of energy must always have been present. On the other hand if the total energy of the world is zero there must have been some agent causing local fluctuations of energy to produce a positive effect in our neighbourhood. Even if energy is not conserved we strongly expect some cause – called God by some – to have created enough of it at a time in the past to lead us to where we are now. But then what caused that cause?

Presented with such difficulties we can turn away from reason and bow our heads to the impossible that faces us. But instead of abdicating from our position of reasoning beings let us stand up to the difficulty as long as we are able, and find out as much as we can about it. Only then can we begin to assess its true nature. We must turn, then, to discover what we can about our universe and its past history.

The first thing that appears clear from observation of the distribution of galaxies in the heavens is that our world has the same appearance in any direction: it is isotropic. This possibility has been tested very carefully in the past few years by seeing if the galaxies are distributed in the same numbers and with the same properties in any direction. No hint of a special orientation has been found; we will suppose it true. If we further assume that we do not occupy a privileged physical position in the world, an idea resisted by many scientists till the early part of this century, but now with very strong support, then we expect the universe to look the same from any position in space. It is, in other words, homogeneous, at least on a large enough scale so that the galaxies appear as points of dust.

We live, then, in a homogeneous, isotropic universe. Where we live in it does not matter; all places are equally attractive. That this is so has very important restrictions on the further properties of the universe, and it has been elevated to the rank of a principle of nature, called the cosmological principle. An extension of this, the perfect cosmological principle, requires further that the universe has the same appearance at all times; when we live is also irrelevant. As Sir Hermann Bondi concisely expressed it: 'geography doesn't matter, and history doesn't matter either'.

It has not proved so easy to obtain experimental evidence for the perfect cosmological principle, but great effort has been made to do so because of the simplicity such a world would present to us. In particular, the universe would then have had no beginning; it has always been here, with exactly the same appearance. Our earlier problems would vanish, though leaving us possibly with a slight feeling of having been cheated because we still cannot answer the question 'but why is it like that?'. However, that should be compensated for by the fact that such a universe has no end either. It is truly immortal, a fitting reward for satisfying the perfect cosmological principle.

Evidence is now at hand which shows quite definitely that the universe cannot be as simple as that. For all the galaxies appear to be moving away from each other with a speed proportional to their distance apart. It is as if they are all embedded in a sheet of rubber which is being stretched steadily ever more and more. This recession of the galaxies was discovered in 1929 by the American astronomer Edwin Hubble and his colleagues, and tested very carefully since then by means of the world's largest optical telescope, the 200-inch Mount Palomar telescope.

To prove this result required independent methods of measuring both the distance and speed of a distant galaxy. Distances were measured in steps, first using nearer galaxies in which certain variable stars could be distinguished. Such stars have known intrinsic luminosity, so that from their apparent brightness their distance can be calculated. It was possible to extend this beyond a million light years by assuming that the brightest stars of a certain type, called supergiants, again had the same intrinsic luminosity, independently of whatever galaxy they resided in. To measure out to distances further than ten million light years, it proved necessary to assume that certain types of galaxies have the same intrinsic brightness, especially in the case of the brightest galaxy of a cluster of galaxies. It is possible by this means to measure to greater galactic distances by such means.

To find the velocity of a galaxy it is necessary to determine how much energy light loses or gains when it is emitted from the galaxy. Radiation loses energy if the galaxy emitting it is moving away from us. There is a corresponding lengthening of the wave-length of the light, which correspondingly looks redder than normal; light from an approaching galaxy has its wavelength shortened, and looks bluer than usual. The shift of discrete wavelengths (arising from emission or absorption by certain chemicals in the stars of the galaxy) can be discovered by comparing the wavelengths of these absorption lines as seen in galactic light to that observed from light produced on earth. This shift, called the Doppler shift, allows the velocity of the object, either towards or away from us, to be calculated.

Hubble's result was that out to a distance of six million light years the recessional velocity of a galaxy away from us is proportional to its distance. It has since been extended to much greater distances by using the Doppler shifts of radio galaxies, whose optical counterparts are often the brightest galaxy in a galactic cluster. The possibility of deviations from Hubble's law at very large distances has been difficult to discover because it is not clear that distant galaxies, being observed as they were a long time ago in the past, would have had the same absolute luminosity, as later, nearer ones.

If we accept Hubble's result we have to conclude that all of the galaxies are rushing away from each other as if they had all been ejected from a common point at some time in the past. The galaxies were then all crowded on top of each other, in a dense blob of matter. This suddenly blew up, spewing its parts in all directions. This is the famous 'big bang' theory of the universe, first put forward by the Belgian prelate Canon Lemaitre.

The matter of the universe is thus supposed to have been a cosmic egg, possibly lying waiting since eternity to be hatched. A perturbation, of unknown cause, made the fertile egg suddenly hatch out our universe, apparently about twelve billion years ago, to mature into the shape we see it today.

There appears an immediate conflict between the idea of a homogeneous, isotropic universe and one which exploded from a point at some instant of time in the past. The centre of expansion certainly seems to be a special place in such a world, which would therefore not seem to be homogeneous. Nor would the galaxies on the edge of the universe be expected to see the same view as those near its centre.

Such a difficulty can easily be resolved by considering models of such expanding universes which satisfy the cosmological principle. One very simple one was suggested as early as 1932 by E. A. Milne. It consisted of a cloud of small particles all emitted from the centre of expansion at the same moment and with all possible speeds below that of light.

From the vantage point of any of the particles all of the others can be seen to be moving away at a speed proportional to their distance. Since the velocity of light is always the same, the edge of the cloud, as seen by any of the particles, is like a balloon expanding at the speed of light. Thus no point is singled out; the expanding dust cloud is homogeneous. This is even true as close to the edge as one likes to go, since there are no particles actually at the edge.

This model cannot be regarded as a realistic model of the universe since it neglects mass and pressure effects, but it does show that there are models of the big bang theory of the universe which are consistent with the cosmological principle. More complicated and realistic models, satisfying Einstein's theory of gravitation, are also possible.

All of these models allow a very useful quantity to be defined which is natural to call the cosmic time. It is the time elapsed since a galaxy of the expanding universe was expelled from the cosmic egg, as measured by a clock moving with the galaxy. Cosmic time is thus a universe-wide sequence of moments, the expansion of the universe acting as a world-wide synchronizing agent at all of its points.

The big bang is a remarkable model of the universe in which change of a very special sort is involved. Yet it fits the Hubble recession of the galaxies so naturally that any rival idea should have great difficulty appearing as attractive. The big-bang model certainly does not satisfy the perfect cosmological principle; the appearance of the universe will depend very strongly on when it is being observed. Because of the elegance of the latter principle a rival to the big-bang theory was suggested about thirty years ago by the British astronomer Sir Fred Hoyle and his colleagues, Sir Hermann Bondi and Thomas Gold. The theory was made consistent with the Hubble expansion by allowing matter to be continually created to fill up the spaces caused by the outward movement of the galaxies. The amount needed was very small, of the order of one hydrogen atom per year in an average 40-storey building, so that it would not be observed locally.

This theory of continuous creation saves the perfect cosmological principle at the expense of the conservation of energy: matter brought into being to fill up the gaps left by the departing galaxies is created from nothing. This gives a very disturbing answer to our earlier question as to where the universe came from: out of nowhere. It has to do so to satisfy the rather mysterious requirement that there is no special time singled out in the world.

Continuous creation would not be necessary in a steady state universe if there were no recession of the galaxies. But the observed increase of the Doppler shift of the galaxies with distance has another very strong argument on its side: why the sky is not bright at night. It is reasonable to suppose that the heavens contain a star in any direction you look at night. But then the sum of all the light coming to the earth will be very high; the dimness of the very distant stars is compensated for by the fact that there are so many more of them than those close up. On this argument we would expect it to be as bright on the earth's surface as on the surface of the sun, which is an average star. It is evidently not.

This difficulty, known as Olber's paradox, can be resolved most naturally by requiring that the light from very distant stars has lost a suitable amount of its energy, this amount increasing with the distance away from us. That loss is precisely what is achieved by the increasing Doppler shift of light from the receding galaxies, and gives strong support to it. For then the light from distant stars loses its importance, and it is dark at night, as we well know.

Very recently strong evidence in favour of the big-bang theory has been discovered by one of the many lucky accidents that help science develop. In 1964, two American scientists, A. Penzias and R. Wilson, were trying to remove the unwanted and unexpectedly large background noise they were picking up on a specially designed horn tuned to radio waves of seven centimetres wavelength. They finally abandoned their attempts and accepted that it was an intrinsic part of the universe. Nearby, another scientist, R. Dicke, when he heard of their result, suggested that what they had detected was the relic of the very energetic radiation which had been excited in the very early stages of the big bang.

Radiation in an enclosure at a certain temperature is called black body radiation, the name black body corresponding to the fact that the enclosure can only be observed without disturbance by making a very small hole in its walls. It then acts as a black body when looked at from outside, trapping any light that falls on the hole and not letting it get out again. The dependence of the energy on the wavelength of such radiation is very characteristic and varies greatly with the black body's temperature. The radiation observed by Penzias and Wilson corresponded to a black body at just over three degrees above absolute zero. More recent measurements have been made at different wavelengths, and all seem to agree on the temperature for this radiation.

This discovery can only reasonably be understood by the big-bang theory of the universe, along the lines suggested by Dicke; it is almost impossible to explain it by means of the continuous creation theory. We are apparently able to see a very clear sign that the universe started off at a special time in the past. It has even proved possible to show that background radiation with a temperature of three degrees is to be expected as the relic of the radiation at far higher temperatures about twelve billion years ago, inside the cosmic egg. In fact the radiation temperature one hundred seconds after the start of the big bang was about one thousand million degrees; since then it has decreased as the inverse square root of the time to its present value as the universe expanded.

This early very hot stage has proved crucial in explaining the origin of the large proportion of helium found in the universe, thought to be roughly 10 per cent of all matter. It has been impossible to understand how such a large amount was created in stellar evolution. For stars get their energy by fusing hydrogen to helium; the estimated total energy produced in this way in our galaxy during its life of ten billion years or so was liberated in the creation of less than one-tenth of the actually observed helium.

In terms of the big-bang theory about one hundred seconds after the bang a certain number of the neutrons and protons rushing around in the hot primordial globe were able to fuse together so that about 10 per cent of the matter became helium. At a later time the temperature had dropped so much that further nuclear reactions were discouraged; the helium was 'frozen in' to the primordial expanding gas, and was later used in building stars by condensation.

It is possible to consider the sort of environment which occurred at even earlier times in the cosmic egg, though with ever greater problems of testing our ideas against established facts. One thing we can immediately recognize is that as we go back in time the three-degree black-body radiation in which we are all bathed increases its temperature due to the increased compression. It does this even faster than the growing energy of the matter in the universe, so that at an early enough stage of the universe radiation was dominant. This is thought to have been the situation earlier than about one hundred thousand years after the cosmic egg hatched.

Before that time the egg consisted of a plasma of charged particles – protons, neutrons and electrons – jostling around and glowing very brightly, being at least at a temperature of one hundred thousand degrees. Going back further in time the temperature increased, becoming hot enough at one billion degrees to cause the protons and neutrons to fuse, as we had noticed before. Even earlier, at one second after the start, it was so hot that pairs of electrons and their anti-matter companions, called positrons (identical with electrons but for an exactly opposite electric charge), were made in abundance by the ever-present light, the temperature then being ten billion degrees. These particles have been recognized to be of great importance in determining the exact amount of helium produced at the later stage, and calculations show that the 10 per cent or so proportion of helium to all matter as observed around us is to be expected.

We can probe back even earlier yet, however. The environment before one second remains very much the same, except heating up, as we retrace our universe's expansion, till we reach back to one ten thousandth of a second. At the corresponding temperature of about one thousand billion degrees a whole new set of events occurred, associated with the excited companions of the nuclear particles – the proton and neutron. These new particles, called hadrons, were first suggested as being the 'glue' holding the neutrons and protons together in the nucleus, in a fashion analogous to the way that the particle of light, the photon, holds the electron and the nucleus together inside the atom.

This suggestion of the existence of 'glue' particles was first made in 1935 by the Japanese scientist Hideki Yukawa. The first of these particles – the pi-meson – was discovered in 1947, and a whole host of its companions, all heavier than the pi-meson – have been discovered since. At temperatures above one thousand billion degrees, and corresponding times earlier than one ten thousandth of a second, these excited particles can be produced copiously. To go back earlier in time, at higher temperatures, it is necessary to know just how many of these particles can be created.

If there are very few new ones made as the temperature increases then the energy of compression at the earlier times will be turned into ever-increasing motion of the particles; the temperature will continue to rise. On the other hand if there are enormous numbers of the new excited particles produced at earlier times then the compression energy could be channelled into creating them, and not making the existing particles move around faster. In this latter case the temperature will not increase. The universe will stay 'warm', at about one thousand million degrees, for times earlier than one thousand billionth of a second.

This second possibility turns out to be most attractive since it agrees very well with known properties of hadronic particles discovered in the last two decades, by using the high energy particle accelerators recently made on various parts of the earth. They have allowed us to discover that increasingly more of these hadrons can be produced as the temperature (or equivalently the energy) is increased. It is by means of these delicate tools made here on earth that we can probe back almost to the very beginning of time. Truly a remarkable piece of detective work!

Such a warm environment, with ever-increasing numbers of heavier hadrons, does not appear possible at the very earliest times, when the natural sizes of the lower mass hadrons are too large to be causal units; light cannot have travelled across them since the time the cosmic egg hatched out. This breakdown of causality means that only smaller hadrons can be created in the cosmic egg at very early times. This restriction (if we believe causality so strongly) becomes operative in the first 'jiffy', a time of length equal to one one hundred thousand billion billionths of a second. Before then the effective size of the universe was so small that the copious production of hadrons inside it was limited by causality restrictions, and so it became ever hotter at earlier times.

To sum up, then, the big bang appears to have started as very hot in the first jiffy, remaining warm from then till one ten thousandth of a second, and then it has been cooling down slowly ever since to come to the three degrees we have now, ten billion years later. But at the earlier stages of the jiffy superspace would have been important, and that is as yet unexplored. So we still don't know precisely how it all began.

One of the important effects of the early stage is the presence of fluctuations in the amount of energy. These variations would produce local aggregations of matter which might grow at a suitable rate to produce objects of galactic size at a later stage. This possibility is being actively investigated at present, and it seems as if such fluctuations might well have occurred at the end of the first jiffy.

If galaxies can arise from expanding aggregations of energy at an early stage of the big bang it is very likely that simultaneously localizations of activity are formed which are so concentrated as to be unable to expand against gravitational self-attraction. They can only produce black holes, which can be of various shapes and sizes. We can put a rough limit on their lowest possible mass if we suppose they cannot be smaller than the size of one of their constituent particles. A rough calculation shows that they must then be heavier than ten millionths of a gramme. They should be moving roughly as fast as average galaxies do, at about five hundred miles per second.

If our earlier estimates of up to 98 per cent of matter being in black holes are correct then some of this vanished matter may be speeding around in quite small black holes instead of in enormous ones at least several times heavier than the sun. A proportion of these 'mini-holes' may possess an electric charge and would behave rather as an energetic proton does when it hits the earth from outer space. A mini-hole would leave a tell-tale track behind it on a photographic plate, very much like a proton does, and such tracks may even have been seen, though not noticed, in photographic plates exposed to cosmic rays at high altitudes.

These energetic mini-holes would require a great deal of matter to slow them up. One of the lightest ones could pass right through the earth, but would be stopped by the sun. It has even been seriously suggested that there are a number of such captured mini-holes which have collected at the sun's centre and are even now affecting it so that its production of certain energetic radiation is not as we expect. That is indeed a pressing problem; scientific experiments have not detected a special form of radiation which should have been reaching earth directly from the centre of the sun. A collection of mini-black-holes there may be playing tricks on us and preventing this radiation from escaping at all. If so black holes are far closer than we thought. What is more the sun would ultimately be swallowed up by its horrific centre, though in a time of at least ten million years – long for us, but far less than its expected age. Even now the sun may be a battle-ground between the dark and the light – and the dark will win in the end.

These mini-holes produced in the very early stages of the universe may also be of use as 'seeds' around which gas condenses in ever-increasing amounts ultimately to form a galaxy. Such seeding will contain in itself its own destruction, as it does in life on earth. For both situations the seed produces life, which then ultimately dies. However, there is a crucial difference, for on earth a new seed is produced to continue the life plan; in the heavens no new seed can ever spring from the collapsed, dead galaxy, come to the end of its visible life. There is but one life between condensation and collapse; the heavenly grave inside an event horizon is final.

Another important problem of the very early stage of the big bang is how protons and neutrons separated from their correspondingly oppositely charged anti-matter companions called anti-protons and anti-neutrons. These anti-particles, as they are called, have the same mass as the original particles, but can annihilate them on meeting, with the production of a great deal of energy in the form of light. In fact anti-protons or anti-neutrons, if they could be kept suitably isolated for a controlled period of time, would make the ultimate weapon, the anti-matter bomb. But the problem in the early stage of the universe is to see how the anti-protons and anti-neutrons (and other anti-hadrons) necessarily floating around in the cosmic egg don't eat up all their companion protons and neutrons before our stage of the expansion is reached.

One of the puzzles of the universe is why there is such a preponderance of matter, at least in our locality, with no trace of naturally occurring anti-matter. It may be that the cosmic egg was itself composed of matter, with very little, if any, anti-matter. But then it is difficult to see why this should be the case. It would be far easier to suppose that the total amount of matter and anti-matter are equal in the universe, and that our local excess of matter arises from a fluctuation or other separation mechanism. That again, is one of the areas of active research at present depending heavily on the properties of the elementary particles which can be discovered in the high energy particle accelerators.

One remarkable thing that the big bang should allow us to do is to measure very accurately our earth's speed with respect to the rest of the universe. For if our globe is moving through the background radiation this should appear hotter in the direction of our movement and cooler in the opposite direction, by the Doppler effect. We should be able to observe this effect as an anisotropy in the background radiation; it should single out our direction of motion.

In fact there is no anisotropy to one part in a thousand, indicating that the earth cannot be moving faster than three hundred kilometres per second. Taking account of the 30 kilometres a second velocity we are moving at with respect to the sun, its 250 kilometres per second around the galactic centre, and our galaxy's 100 kilometres a second motion in its local group tells us a lot about what all these motions can still only add up to. There are already hints that we are on the verge of observing a special direction in the background radiation; we are about to have our first experience of the total effect of the whole universe on ourselves.

There is a basic difficulty about how the universe became so elegantly homogeneous. At the very early stages of the big bang it is very likely that some parts of the universe were not visible to each other till quite late in the proceedings. The magnificent isotropy of the background radiation is then hard to explain, because that arising from two different directions more than about 30° apart may have been previously scattered by parts of the early universe which did not know about each other's existence. How could they then send us the same amount of radiation for our delight?

It turns out that this problem comes from the existence of another sort of horizon than the event horizon we have been so concerned about up to now. It may be visualized if we picture ourselves as flatlanders again, living on the surface of an expanding sheet of rubber, say on a balloon. The big bang will then correspond to the initial puff of air starting the expansion of our balloon world. Each galaxy can be pictured as a speck fixed to the surface of the balloon. The recession of the galaxies naturally arises from the expansion of the balloon.

Light emitted from a galaxy crawls along the surface of the balloon to us. At the same time we are receding as the balloon expands, and it may be possible that the balloon expands at such a rate that the light never gets to us. As the British astrophysicist Sir Arthur Eddington has put it, the light is then like a runner on an expanding track, with the winning post, us, receding forever from him. The light from some galaxies may never reach us, so they are never visible to us. We may call our event horizon precisely all those events which we will never just be able to see. Those things happening outside it are forever beyond our possible powers of observation.

The expanding universe we live in gives us an event horizon, when each galaxy in its flight from us reaches the speed of light. That happens for us now when it is about twelve billion light years away. We have not yet seen out to that ultimate distance but are already as close as three-quarters of our way there. It turns out that if we do truly see such a horizon then we must be prepared for all the galaxies, even our nearest neighbours, ultimately to pass beyond it and vanish from our sight. However we are not sure that one exists; if the universe's expansion slows down sufficiently there will not be one, so such a fate need not worry us. However, there may be worse in store for us than loneliness in that case, as we will see shortly.

The event horizon around a black hole forms part of our total event horizon. We can expect that particles on the surface of the black hole are falling in at the speed of light, so that we can never see any light they emit to us; once inside they are falling even faster. This picture seems to violate the sacrosanct rule that 'you cannot go faster than light', but that only holds over very small distances in a highly curved space. So a black-hole event horizon is just part of that total surface surrounding us and shutting us off from observing anything outside it.

The other sort of horizon occurs when the very first light from a galaxy finally reaches us after crawling over the surface of our expanding balloon. That galaxy then suddenly appears to us, being created in all its glory in the firmament. We can divide the galaxies at any one time into two classes, those visible to us and those not; the dividing line is called the particle horizon.

There are various possible ways in which the universe could evolve after the big bang so that there are non-trivial particle horizons. This will mean that parts of the early universe cannot influence each other satisfactorily to explain the observed isotropy of the universe. The absence of appreciable particle horizons imposes strong conditions on any model of the early stages of the universe. Various schemes have been recently suggested which do avoid this difficulty, but a great deal more understanding is necessary. Whatever that will lead to we do not know, but we have already cast aside the first of the many veils over the mystery of the cosmos. And undoubtedly we can remove many more in the future.

We have to accept, then, that about ten billion years or so ago the universe was far smaller and hotter than it is now. Indeed if we trace backwards in time we will expect to reach a point when all matter and energy were condensed to a point, and time, space and very existence itself had no meaning. Such a possibility would seem to be abhorrent to us, since we cannot hope to discuss anything about the nature of the world before then. We appear finally to have met the impossible. We can go back no further.

One approach to this difficulty is to suggest that time itself is not a linear phenomenon. As the universe expands from its singular point the first moments of time are of the utmost importance. The changes that occur then are very great, and there may even be an infinite number of them. Time passes very, very slowly for the first few billionths of a second, because of the great activity.

A little later the amount of action is reduced and we could even recognize the beginning of the existence of the elementary particle which, comprising the neutron and proton, and their companions, appear to constitute the world we live in now. Before that time those fundamental particles had been broken down into their constituents, before that those constituents into their constituents, and so on. As activity occurs time passes at a corresponding rate, the universe expands ever more, and its temperature drops. The rest of the story until ourselves is lengthy and complicated, but comprehensible in principle.

On this picture of time being determined by the amount of activity in the universe there was no beginning to the world; it already existed. What we conceive of now as the passage of the first billionth of a second would have seemed to us to have taken an eternity if we could have experienced it in all its glory. So we have again a model which has no beginning for the universe: in the beginning there was no beginning. And we are here because we are here because we are here.

This picture is an attractive one if we find, in our search for the ultimate constituents of nature, that they always have their own constituents, and they theirs, *ad infinitum*. For then there would always have been new activity occurring, however close to the very beginning of the big bang we could probe; this would then correspond to new time. This activity would be due to the appearance of the new constituents as time was pushed ever farther backwards to its beginning. It is interesting to notice that according to this approach the attempt to understand the world in the very small also sheds an enormous amount of light on the whole universe.

It is possible to go even further than linking the microscopic and macroscopic worlds. In Chapter 3 we introduced the idea of a self-consistent world, one which 'held itself up by its own bootstraps'. There have been reasonably successful attempts to construct a model for such a world by showing how the elementary particles can suitably engender themselves.

The number of particles of a certain mass will then increase very rapidly with the mass. By the argument we gave a little earlier the universe, after the first jiffy, will remain at roughly the same temperature. This will occur till there has been so much expansion that the elementary particles are 'frozen in' at about a ten thousandth of a second. So our whole universe could be constructed initially out of self-creating entities. It would then be truly relative: nothing would be fundamental, everything would create and in its turn be created out of everything else.

On the other hand, if there are only a limited number of constituents of matter which are truly fundamental, with no further constituents themselves, then such a picture of ever greater activity at earlier times closer to the initial explosion loses its credibility. In that case we can suppose that the cosmic egg had a smallest natural size, when all the ultimate constituents of matter were all squeezed up against each other as tightly as they could be. This minimum egg might have been sitting since eternity when suddenly, due to some unknown perturbation, it expanded and became as we now see it. Such a picture has the very great difficulty of having to explain why the egg wanted an infinite time to hatch out. Indeed if it had waited so long it would seem more likely that it would never hatch but still be waiting. Only if it had been created a finite time ago with a finite lifetime does one expect such a hatching ever to have been able to take place.

So we have to ask ourselves if it is possible to think of a model of the universe which would explain how the cosmic egg came into existence at a finite time in the past. We know that it is not enough to say it just arrived; the question is from where. There are basically two possibilities for this. The first is from a rotating black hole in the previous universe. We can suppose that the fate of a spinning collapsar after reaching its final state of collapse is to reappear in a completely differently universe as a spinning white hole.

Such a white hole can be visualized as developing in exactly the reverse to a collapsar; it would appear as would a film taken of the collapse of a rotating star played backwards. Yet we have already noticed that the collapsar would have to arise from a previous universe and we are still left with the unanswered question of where that collapse came from. If we say it itself came from a previous universe then we see that the sequence of universes need never end. We have already regarded this as unsatisfactory, but we should not let our personal preferences distract us; it may be the truth.

There is information left behind by the background radiation which already indicates that the universe has almost no rotation. Before the three-degree background was discovered the best that one could say was that the universe rotated no faster than once every hundred million years. That seems a long time, but means in practice that since our own galaxy was created the universe would have rotated about one hundred times. This figure can be reduced a million-fold if we use the observed level of isotropy of the background radiation; since our galaxy came into existence the universe can only have spun round one ten-thousandth of a turn, a negligible amount.

If the universe did arise from the escape of a rotating collapsar from its own universe it only just managed to; the collapsar was close to being a stationary one. That must have been so in any case in order that the initial phases of the big bang were energetic enough. However, it seems difficult to give any reasonable value for the spin of the rotating black hole which produced the cosmic egg other than zero or the maximum value to avoid a naked singularity. But the latter value is certainly ruled out by measurements, so the former is to be expected. As remarked above it appears to be very close to the facts.

We have to return to a non-spinning cosmic egg. Can it have arisen from the collapse of a previously expanded universe which we can still identify as our own? We might suppose that a similar fate awaits the material at the end point of its collapse in a stationary black hole as it does in a spinning one. In other words it bubbles out into another universe. It certainly cannot re-expand into its original world without some pretty drastic assumptions, but it might go through the gates of superspace to expand into the elsewhere; just as well it can have come from the elsewhere to become us.

This mechanism of travelling through the world of many possible geometries to produce us also has its inherent difficulties. The previous world would be destroyed at the portals of superspace; to enter into this new space the size of the old universe would have to be immensely small. In order to be able to follow the system through its development we would have to know something of the laws which govern matter, if it can still be called matter, at these enormous densities.

Undoubtedly the form these laws will take will be far removed from anything we know or even glimpse of today. But there must at least be some rules, for even if there were complete chaos governing motion in superspace, then at least that would be a law. In fact the very idea superspace itself is based on the idea of probability, so we do expect some form of probability statement about the evolution of the world. But here we come to the crux of the difficulty about this approach to the beginning of the universe. For probability applied to the whole universe has a very hollow sound; it even appears as if we cannot even use the concept at all.

To say that there is a 50 per cent chance of tossing a head on a coin is really to say that out of a large number of tossed coins, or the same one tossed repeatedly, about half of them have the head uppermost. As the number of tosses is increased the proportion of those with a head becomes closer and closer to 50 per cent. We cannot say anything about the result of a single toss of a coin, so that the probability concept only applies to a large number of cases of an event.

How can we interpret the statement that the universe is in such and such a configuration with a certain probability? For example, that it has a billionth of a per cent probability of the earth being only half its present size, a feature of some of the possible worlds we could experience in superspace? For this statement to make sense we (or at least someone) should have to be able to investigate a plethora of universes to be able to determine the proportion of those for which the earth is only half its present size.

By definition there is one universe. It is unique, and contains all that is the case. Or at least it ought to do, since we have already noticed that if it does not then whatever had been left out in its first construction can now be added in to get the whole lot. So that for the unique universe the probability idea, and so superspace, cannot be possible. If our universe is the whole world – all that there is – then it cannot have come via the cosmic egg through superspace. It can only have come in that manner if it is not all that exists; there must be some elsewhere. In fact, there must be a large universe, in order that the probability notion is sensible. A cosmic observer can then determine the probability of the number of such elsewhere containing many copies of our earth being half its usual size, say, by looking at all these copies of the universe as they develop from their cosmic egg stage.

This multiplication of universes is forced on us by the need to investigate the very early stage of the universe; when it was enormously compressed probabilistic effects would be expected to be important in the same way that they are in the world of the very small, that is to say in the atom and even more so in the nucleus. But there is a further reason that probabilistic notions should play a role in the universe at large. That arises from our supposition that we can build up the properties of any object knowing those of its constituents. We are not necessarily denying that the whole may be greater than the sum of its parts, since the way these parts interact may determine gross effects of a different nature from those for the constituents. All the same, we suppose that only the constituents are crucial.

If we apply this to the whole universe then we expect to be able to describe its behaviour in terms of its constituent elementary particles the proton, neutron, and their companions. These obey laws which are decidedly probabilistic, as recent very careful experiments have shown; there are no 'hidden variables' in terms of which the probabilistic features can be described as arising due to ignorance. But then the universe itself should have such probabilistic features. Because this does not make sense without a plethora of universes, as I have argued above, then we cannot be constructing the whole universe out of these probabilistic elementary particles.

This is a difficulty which does not only arise from the early stage of the universe but also at any later time. It is, however, expected to be decidedly more important for the whole world in the beginning of the big bang than at later stages, since in the very early stage the particles will be in a state which is expected to be highly correlated. Later the particles seem to have far less effect on each other, except on those very close.

How can we resolve this difficulty? It arises both in the case we have just been considering, that of a cosmic egg which arrived on the scene only about ten billion years ago, and also in the earlier one when the egg had always been there because time was stretched out to eternity at the beginning of the bang. We must accept either that there is a large multiplicity of universes like our own or that alternatively the probability notion breaks down when we go to a deep enough level of the constituents of the universe.

The first possibility we had dismissed earlier because there can only be one totality of things. Or so it would seem at first glance. But if they are separated by event horizons, each from the other, such as a pair of distant black holes would be, we may legitimately regard each as the unique universe to its inhabitants. They would not affect each other, and would indeed be superfluous as far as the other universes were concerned. But yet they would be found necessary to be assumed to exist in order that probabilistic notions would apply. If it were found that probability is essential in understanding our own universe at the deepest level then such unobservable universes would have to be assumed to exist. They would only affect us by the laws of chance governing all the universes.

The second situation, that probability breaks down if we delve deep enough into the fundamental particles, can only be tested by experiment. If it is so then our present understanding of the laws of physics is in for a shock. But then we are very likely in for many shocks over the coming centuries as we explore the universe ever more closely. Without probability we lose the chance of a cosmic egg being laid by a collapsing non-rotating star through superspace, since the latter can no longer exist. It may be laid in a different manner but we have no inkling of what that could be.

There is one other possibility which should be mentioned, though its conclusions are so powerful that it is to be treated with great caution as hiding a defect somewhere. If we wish to find out the exact probability at a given time of tossing a head with a coin we need to have an infinite number of copies of the coin. For otherwise we may only be able to toss up fifty similar coins, or five hundred; the proportion of coins coming down head cannot give the exact probability of tossing a head, though the larger number should give a more accurate result.

In the same way if we wish to get accurate probability results for measurements on any part of a homogeneous, isotropic universe at any time, then it must be infinite in extent; any part will only then be duplicated infinitely often and so may be accessible to observation. A probabilistic view of the universe requires it to be infinite; a closed universe cannot be probabilistic. If we believe the probabilistic microworld is a true substructure for the universe as a whole then it follows that we must live in an infinite universe. This is indeed a powerful result, but one requiring a very big jump. Undoubtedly, we will have to look a little more before we can make that leap!

In conclusion it is most likely that the universe commenced its history as a highly compressed cosmic egg which suddenly expanded about ten billion years ago. This time span may not be that experienced by the universe itself; this latter proper time could have been stretched to infinity at the beginning of the big bang. Thus the universe had no beginning in this account. Such an idea is only sensible if there is a never-ending sequence of constituents of constituents, etc. etc., of the matter in the universe. At present we certainly do not know that, if we ever will.

If there is only a finite chain of constituents of constituents, etc., then the cosmic egg was ready to explode into the big bang a finite proper time ago. In that case its only origin can have been from the non-rotating collapse of another universe through superspace; this is only feasible if there are a plethora of other worlds only linked to ours by probability notions, or possibly if the universe is infinite in extent.

More succinctly in the beginning of our universe there was either no beginning or there was the end of a previous universe. Naturally the former of these two possibilities can only be investigated by looking in closer and closer detail at the chain of constituents of matter leading, we hope, ever downwards. But the second possibility can be further considered by looking at the end of our own universe and asking what it could be like. That we will do now.

10 The Inevitable End

The imminent end of the world has been prophesied many times in man's history, and even believed to such an extent that true believers would give away their property and belongings to prepare for it completely. There was one religious sect whose prophetess caused her followers to shave their heads except for a handful of hair on the crown, so that they could be snatched from the ground by angels to save them from the destruction to come. She made them build a tall wooden scaffolding and had them stand on it so as to aid the angels in their appointed task. The only effect on the world of its own destruction was that the scaffolding collapsed due, it was suggested by some non-believers, to perfectly natural causes. It was, indeed, the end, but only for the number of the faithful who perished in the disaster.

In this and all similar cases the end of the world is only the end of the world as we know it, not a total destruction of it into nothingness. It is difficult for us to comprehend so drastic a fate as complete cessation of all being. Even our own death is difficult to visualize. Indeed how can we imagine a time when we cease to imagine? We can, however, picture other people still living and going about their daily tasks. Yet if there is no universe even that is impossible – there is just nothing at all to think about in such a situation. So we hang on to our idea that the universe, or some aspects of it, will continue existing for ever.

We may well be completely wrong: the end of the world may be precisely that, with nothing to follow it. Even if it is as final as that we can still attempt to understand how such an end is to be met – what the physical conditions will be before the vanishing trick. And in any case we should try to push our picture of the universe's future to the ultimate possible. We can do this by using our present understanding of the world's past and present to predict its most probable future. It might even be possible to plot the future course of events to their ultimate conclusion, and so dispense with the need to have the universe disappear into thin air.

We can go beyond vague philosophizing if we recall that there is experimental evidence against the idea of continuous creation of matter, and in favour of the big bang theory for the origin of the universe. So at least we know that there is no continued creation of matter out of nothing. This evidence does not, of course, prove that there could not have been a fantastic production of something out of nothing, when the big bang itself occurred. But the creation of the whole universe out of nothing, in one fell swoop, so to say, appears to be impossible, even more so than its disappearance. It just cannot be achieved because there would be no way for any self-creative process to inform itself to start creating the universe and itself as part of it.

There is, on closer investigation, a similar problem for the universe to cease to exist entirely. For whatever mechanism would achieve this has to cause itself to cease to exist. In the process of its own destruction it will fail to function properly to achieve the final annihilation. It is as if the doomsday machine has to crush itself and the world to a very, very fine powder. But at the split instant of the explosion the machine has blown itself apart and so cannot continue crushing the world, or itself, to ever finer and finer pieces. At least part of the world would survive such a holocaust.

So let us reject the complete vanishing trick for the world's end, and consider what else might happen. We know that the galaxies are receding from each other ever faster; this is the Hubble recession for which there is strong experimental evidence. There are two possible fates in store for such an expanding universe: either the galaxies continue their flight from each other for ever and get farther and farther apart, or alternatively they slow down and ultimately stop, eventually to be pulled back together by their mutual gravitational attraction at an ever-increasing speed. They will ultimately collide together in a highly collapsed object undergoing just the reverse of the big bang. They would make a universal collapsar.

The first of these possibilities, that of the ultimate escape of the galaxies from each other, would be one which we can only face with despair. For as the distance of a galaxy from our own increased so would its speed, till ultimately it would be travelling faster than the speed of light with respect to us. This would not seem to be allowed by theory of special relativity, but in fact it is for distant objects in a curved space. This curvature causes strong distortions of space and time, so that observers at far-removed places have very different rates of passage of time from that nearby (something we have already noticed near the surface of a black hole).

As galaxies reach the speed of light in their flight from us we will cease to see them. They will disappear through our event horizon, as we described in the last chapter. So one by one the more distant galaxies will disappear from our view; later on nearer galaxies will vanish. Ultimately we might be completely isolated with only our local cluster of galaxies around us. And even if nearer galaxies do not flee from us at faster than the speed of light they will still recede from us so that ultimately we will not be able to observe them satisfactorily. The astronomers of the future would have had to preserve the record of the heavens by taking a sufficiently large number of photographs while distant galaxies are still visible. One can even envisage future communities in which there are ritual replays of such records, with appropriate ceremony to mark the occasion.

The time taken to reach a noticeable loss of galaxies from around us is long. For example, the time taken for galaxies to become twice as distant from each other as they are now, is at least the present age of the universe, but it still will be reached. For time will continue forever in such a future; there would never be an end to it. It could well slow down due to the ever-decreasing activity which would occur; the local galaxies would age till eventually all of their contents had evolved to the final stage of white dwarfs, neutron stars and black holes.

These latter do present a problem, however, for such an

immortal dilute universe. They provide local centres of collapse which could ultimately absorb all the matter round them. Sooner or later that would, indeed, be the fate of all white dwarfs and neutron stars; they would be swallowed by the black holes with emission of gravitational radiation. Ultimately our galaxy and the others of the local group of galaxies to which the Milky Way belongs, would be completely swallowed up; we would all descend into our central black hole.

Such a fate is even further off than the disappearance of the galaxies. But as far as we understand black holes, and more generally the nature of space and time, such a fate is inevitable. However hard we struggle by preparing to leave our galaxy we would undoubtedly be captured, if not in our own galaxy, then in another to which we might have fled. For we expect that there would be at least one black hole formed in each galaxy, if not many more. For example it is thought that at least seven candidates for collapsars are born in our own galaxy every year, and we expect other galaxies to be similar to it sooner or later. We already have a very good candidate for one in our galaxy, the X-ray binary star Cygnus X-1; hints of a further one in the Southern Hemisphere have very recently been found.

Such a fate strikes hard at all human aspirations. We accept mortality, but hope that our strivings for the future will be preserved in some form or other. By this we hope to attain some proportion of the immortality for which we long; we especially aspire to this through our children. We have to accept that such immortality is impossible in such a future. Billions of years will pass, but the black holes will get us in the end. And then we know that our future is uncertain to an extreme.

There appear two ways of escape from such a grim prognostication. One is to flee to regions remote from any black holes and try to forge a living out of the barren surroundings. This would be immensely difficult, and can be regarded as well-nigh impossible because of the nature of intergalactic space. There would be no reasonable energy sources for us to draw on far from galaxies; we would face death by freezing if we stayed safely away from the dangerous black holes in galaxies.

That fate could overtake us in any case if we stayed in a galaxy in which there were, miraculously enough, no black holes. For all the stars would ultimately have evolved to neutron stars or black dwarfs and could give us no further energy. We would have to build our own stars to warm ourselves; even then we would ultimately use up all available matter as nuclear fuel. The search for safe galaxies would only appear to be putting off the evil day, but never completely retarding it.

The other way of escape is even more difficult to envisage. It is to use the rotating black hole as a means of fleeing from the threatening universe into another world which hopefully would be less dangerous. This has the basic danger that we would never know, until it was possibly too late, if such a trip would succeed. Not till we were well inside the event horizon would we have any hope of determining if such an escape hatch to the other world were open or not. Even if it were we would still be uncertain till even later what the exact size and nature of this gateway to the new world might be. We could theorize and predict to our heart's content, but we would never really know if it would succeed till the jump was attempted. Could we trust our lives on that?

Even if the world appeared so threatening that we would feel impelled to leap before we could look we might still be going out of the frying pan into the fire. The new world could have an even more threatening collection of black holes in it than the one we came from. Certainly we would not know if that were the case until we had gone far beyond the point of no return. Here again we would have to take a gamble, but one with a far higher uncertainty. For though we could not be absolutely certain about the truth of our predictions on the nature of the interior of a rotating black hole, we could still have reasonable faith in them; we have far less assurance about the other world which can be reached through the black hole.

The fate of intelligent beings near any star in any galaxy of the universe appears bleak, even if the universe expands indefinitely. It is possible to assess whether or not this occurs by making sufficient measurements of the total amount of energy distributed throughout the universe. For this energy is equivalent to mass by Einstein's famous equation '$E=mc^2$'; gravitational attraction will cause any distribution of energy to contract to a smaller size.

On a large scale it is a reasonable approximation to assume that the various forms of energy in the universe – matter in galaxies, intergalactic hydrogen, starlight, and various other manifestations – are spread out smoothly. It is then possible to calculate the kinetic energy – the energy stored in motion – of the expanding smoothed-out universe. We can also estimate the gravitational energy which could be released if the universe collapsed to a point. The total energy – kinetic plus gravitational – gives a critical estimate of the future fate of the universe. If that total is positive then the universe can go on expanding indefinitely, doing so with constant speed after a long enough time.

It is only if the total energy of the universe is negative that such expansion will slow down, as time proceeds, at such a rate that it ultimately halts and reverses; the universe begins to collapse, as was mentioned earlier, and falls back on itself to produce a state of infinitely high density. The intermediate case arises when the total energy of the world is zero, and then the galaxies recede at an ever slower rate, though getting ever farther apart. The fate of isolation will then undertake the member galaxies as it did in the case of positive energy.

The crucial dividing line between ever farther isolation or a return to the singular totally collapsed state occurs at a definite value for the average density of energy throughout space. If there is less than this critical value of energy distributed in the universe then the gravitational attraction between the parts of the universe are too weak to overcome the impulse to expand imparted by the initial 'big bang'; the galactic constituents of the universe will continue hurrying away from each other in spite of their attraction to each other and the other forms of energy. On the other hand if there is more energy than that special value then the initial energy of recession is ultimately conquered by gravity and universal collapse ensues.

It is evidently of great importance to find both this crucial energy value and the actual numerical amount of energy residing in the world. The former of these values can be found from the speed of recession of the galaxies, and turns out to correspond to a uniform distribution throughout space of one atom of hydrogen in twenty-five thousand cubic miles. This amount appears to be very little, but it must be realized that it is, in fact, ten times more than would arise if all the matter in the galaxies were distributed uniformly through space in the form of hydrogen. So it is necessary to search very carefully for other forms of matter or energy before we can expect the universe to collapse back on itself.

There are numerous modes of energy which require investigation before the future is clear. Some of these – especially the energy of starlight – are so small as to be negligible; that arising from interstellar magnetic fields, from cosmic-ray particles or from the background radiation, can similarly be neglected. There are two further forms of energy, however, which make the answer very uncertain. One of these is that of invisible matter in the form of collapsed stars or even collapsed galaxies. We mentioned earlier that there might be even fifty times or more collapsed matter than in the normal state, so in that case the universe would undoubtedly collapse some time in the future. It is difficult to give any precise figure at present for the density of invisible matter. The conservative estimate of 50 per cent would not be enough to bring the universe together again, and at least 90 per cent would be necessary; this was the amount mentioned earlier as needed to explain the formation of heavy elements in the galaxy.

The other form of matter which it has proved very difficult to observe is in the form of intergalactic hydrogen. It has only been possible to put an upper limit on its presence, and that seems too meagre to be attractive enough to re-collapse the universe. Yet there are severe difficulties in interpreting the observations that have been made so far, so that there still may be more than the critical amount of hydrogen between the galaxies.

There is also radiation energy in much more elusive forms which has to be taken into account before the final answer can be given. One of these types is gravitational radiation, the gravitational analogue of light and other forms of electro-magnetic radiation. This has proved very hard to observe, though we mentioned earlier its possible discovery by Joseph Weber. Yet it is very hard to measure how much of it there is throughout the universe, especially over the whole range of frequencies. In particular there may well be a large amount of it left over from the big bang, being the gravitational version of the background radiation. There is also an elusive equivalent to electromagnetic and gravitational radiation which is that of neutrinos. These are particles with no mass or charge but an equal spin to that of the electron. They were introduced to explain radioactivity, and later discovered. Un-doubtedly there will be a relic of the original neutrino energy flung from the big bang, but yet again that will prove very difficult to see.

Direct observation of the speed of expansion of the most distant galaxies has shown that some of them may already be slowing down in their flight from us. If that is so then there must be enough matter around to cause such a retardation. However, the observations are not accurate enough to be cer-tain about this.

The situation at present then, is that we don't yet know if the universe will continue to expand or will ultimately stop and collapse. We have already considered the fate in store in the former of these possibilities; let us now turn to discuss the other case. We will find that some curious effects can occur in the collapse phase long before high densities are reached to put intelligent life in danger.

It is useful to remark here that the condition of high enough density for ultimate collapse is exactly that for the universe as a whole to be inside its event horizon. In other words we live in a black-hole universe. If that occurs we must regard the space of which the world consists as being so strongly distorted by the large concentration of matter in it that it is curved back on itself. Such a space is the three-dimensional analogue of the curved surface of a sphere. This latter has a finite area, but is unlimited in extent in the sense that it is never possible to reach a boundary of the space.

A two-dimensional being – a flatlander – living on a sphere can never experience anything outside it; his world is limited to the spherical curved surface. If our universe consists of a space which is closed on itself we will be the three-dimensional analogues of the flatlanders. We will never be able to leave our closed universe; it is impossible to signal outside it by any means. However, we cannot be sure that nothing can come in from outside, though it could only do that if it came through a higher dimensional space. This is analogous to objects coming in from outside the flatlanders' world via the third dimension (which they do not know about). We will never know for certain if there are any further spacial dimensions beyond our current three; it would only be through the unexpected appearance or disappearance of objects from our three-dimensional world that we might have an inkling of such extra dimensions. Since there is no impelling need to introduce them we will not do so here.

So if we live in a closed universe we can say that we are sealed off from elsewhere. In fact, for such a universe there will be no elsewhere. We can certainly tell if our universe is closed by some peculiar properties it will have. These can be seen by analogy with the flatlanders' spherical surface. If we draw circles of given latitude on the earth, starting with those at the North Pole, their circumference increases as we recede from the North Pole until the equator is reached, and then it decreases till finally it is zero at the South Pole. However, the corresponding area enclosed between the circle and the North Pole increases steadily all the time.

Similarly in a close three-dimensional universe a set of spheres of ever-larger radii, with a fixed origin, will have their surface area initially increasing, but finally decreasing to zero, while their volume will steadily increase to a certain maximum value, the total volume of the space. In principle we should be able to make such area and volume measurements, and so determine if our universe is closed or not. In practice that is obviously unfeasible. However one property of a closed universe is amusing to note: a light ray sent from a given point will ultimately return to that point. So if you look carefully enough in a closed universe you should be able to see the back of your head! This was even taken seriously some years ago when a galaxy emitting strong radio waves was discovered in exactly the opposite direction in the southern hemisphere to one earlier discovered in the northern one: one was expected to be the back of the first. That naïve possibility has to be rejected due to the recession of the galaxies, as well as other effects.

Besides being able to see the back of one's head if one looks hard enough, a closed universe has some very interesting features as far as the compression stage is concerned. The most bizarre of these is in the experience of time, and especially its direction. We have seen how the interior of a spinning black hole may be used as a time machine, and one can even use any exterior region if a naked singularity can be constructed. But there is a natural 'arrow of time' which is related to the increase in probability in a particular process.

For example if the particles of a gas were in a box with a top on it, and the top was then removed, the particles would ultimately disperse out of the box and wander around in space; it is highly unlikely that at a later time they would all spontaneously find their way back into the box again. There is a slight possibility that they might indeed do that, but it is vanishingly small. We expect the world to evolve in a manner in which more likely events occur; this determines a direction of time.

We can relate this sequential ordering of natural events to the expansion of the universe by noting that the world is a cold one, absorbing radiation. If we shine a flashlight, the radiation it emits is sucked up by the hungry universe; this gives a definite sequence of before-after with which we are familiar. The arrow of time given by the increase of probability can now be equated to that given by the expansion of the universe.

In a contracting universe we would then expect that as collapse proceeded less likely events would occur – for example the gas particles would move back into the box from which they had escaped in the expanding phase. The galaxies would move closer together with their constituent stars absorbing the radiation they had originally emitted before contraction had set in, and ultimately disintegrate into the primeval hydrogen gas. In fact the further development of the world would appear exactly as if a film of the expanding phase had been played backwards. People would arise from their graves, grow younger and eventually be unborn – going much further than needed by the believers in immortality as to what will happen on the day of Judgement.

This picture of humans ungrowing and finally being unborn is a disturbing one, but it can be completely avoided by considering time as reversed, as experienced by intelligent beings in this contracting phase of the universe. In fact that would be the only way that time could be experienced by humans in such a world, since we only notice the 'psychological' direction of time. That is determined by the expansion of memory, and will correspond to the reverse of the physical time in the contracting universe.

In any case there is no real reason to expect processes such as stellar evolution to run backwards in the contracting phase. This activity should continue till nuclear fuel is exhausted, being finally stopped by the increasing light in the sky coming from far-off galaxies which are now coming closer together. The inward motion of these galaxies will cause the star-light they emit to gain energy (be blue-shifted) and so give ever-larger energy contributions to the sky. Sooner or later this increase will cause the night sky to be as bright as it is in the day, distant galaxies contributing as much energy to the earth as is done by our own sun.

Ultimately, the earth and all the stars would vaporize under this glare of radiation, all life in any form having been annihilated well before then. The universe will continue its inexorable collapse till all structure existing previously – planets, stars, clusters of stars, galaxies, and clusters of galaxies – is wiped away in the seething cauldron of matter and radiation at ever-higher temperatures and densities. Nor is there any end to the compression, as the constituents of the constituents are stripped bare, and then their constituents, and so on.

This destructive collapse of the whole universe will be as

inexorable as that at the centre of a collapsar. Nor can any pressure exerted by the ever-compressed material help evade the final destruction, since such pressure would only add further energy to the universe. Because of this increase the collapse would only be speeded up. There is no way out.

What happens finally? We can evade this question by replying that there need be no end to the unravelling of the constituents of matter: any constituent of matter always has its own constituents. As in the beginning, when there was no beginning but always activity of ravelling, fusion of constituents, so in the end there will be no end, since again activity will always continue with the ever-possible unravelling of further constituents.

This picture of the universe with no beginning and no end is attractive for its simplicity. Yet we have indications that there are natural sizes below which matter may find difficulty in being compressed. The natural unit of length is very small, being a million billion billion billionths of a centimetre, but this arises as a natural length below which the force of gravity could even be ineffective. Though there is no evidence as yet, this is a possibility which should be borne in mind when assessing the possible validity of various present theories.

The other alternative to an end without end is a re-expansion after compression, possibly to a size dictated by the natural unit of length I quoted above. In that case the whole story would start all over again – expansion, slowing down, maximum expansion point finally being reached, contraction, ultimately with destruction of the significant features we see around us. Such a re-cycling would continue indefinitely, and may have already done so in the past.

It is certain that in either of these endings – end without end or re-cycling – humanity and all intelligent life will not survive the phase of high compression. Even the fundamental particles of which we are composed – the electron proton and neutron, along with their excited unstable companions – will very likely be decimated as well, though long after ourselves. There appears to be no escape for us in either situation after a certain length of time, at least in our present universe. We will never attain immortality in a closed universe.

All possibilities point to a difficult time ahead for intelligent life in the universe. The best bet appears to be an expanding universe in which our fate is first to be frozen to death and then to be trapped by the final black holes. Whilst we will still be annihilated in the end, we would expect to put off the evil day much longer than if we live in a closed universe. It is important, then, to find what our fate will actually be. But beyond that we have to ask how we can face up to it. Will we have to admit final defeat in the face of the inevitable?

11 The Immortals

We are not alone in the universe. It is certain that there is at least one planet with intelligent life on it circling one of the billions of stars in one of the billions of galaxies somewhere else in the heavens. It is very likely that there are many life-supporting planets even in our own galaxy, with a proportion of them bearing intelligent beings. Such life may be very different from the varieties which have developed on our globe, and could have a very different view of the universe from our own. The mysteries facing other intelligences in the universe may well be different from those we have tried to grapple with so far. It will help us to get a better perspective of the puzzles of the world if we try to see it through others' eyes. At the same time we will be attacking one of the most personal problems we are presented with: the nature of our own intelligence.

We are faced with an even more basic question when we turn to consider life itself. What does it mean to say that an object is alive? Life on earth has various characteristics which we could use to test for it elsewhere. The most important are grouped together in the four F's: feeding, filling out, fighting and reproduction. All earthly living organisms take in food in order to supply energy and also to provide material for growth and replacement. They grow from within, with the replacement of worn-out parts or the development of new ones. Then there is reaction against the environment to preserve the identity of the being. Finally, there is the need to reproduce, to preserve identity of the species.

The four F's are properties of all things living on earth. No inanimate object possesses all of them, however beautifully organized it is. A flame consumes oxygen to give it the energy to keep alight, it can grow as it burns, it reacts to its environment – to a wind, say, and it can produce new flames. But these descendants of the old flame may be very different from their parent. It is the lack of identical reproduction which leads us to deny a flame the right to live. Similarly a crystal can feed on the surrounding solution – so enlarging itself, can modify itself if there is a change in the temperature becoming smaller if the solution it is in is heated, but it cannot reproduce its own kind.

Reproduction may not be a crucial property of life elsewhere. There has been very little discussion of such a possibility, but we should be prepared for it. One form of non-regenerating 'life' was suggested by Sir Fred Hoyle in his science fiction novel *The Black Cloud*. This cloud was composed of an aggregation of cosmic dust with various self-organizing electrical powers raised to such a high level as to create a motherless and fatherless yet sentient, self-aware being. Without birth there need be no death. Provided the being had suitable energy supplies to keep its organized state at a complex enough level it could exist forever till the end of the universe. The black cloud achieved this by surrounding a star and absorbing the radiant energy from it.

Life without reproduction is difficult for us to comprehend. We ask how it would be possible for a highly complicated system, capable of life-like behaviour, to develop without the constant attack and retreat of the battle for the survival of the fittest. Yet there is no reason in principle why an organism could not develop by continual modification to its internal organization, so as to keep pace with changes in its environment. The competition for survival would then be not between various types of life for the best 'place in the sun', but between a single organism against a dynamic but inanimate universe trying to bring the animal into the same state of chaos as its surroundings.

We have here a suitable criterion for life in the universe. An object passes the test if it involves a state of organization which can preserve itself in roughly the same form in spite of important changes taking place around it. By the idea of organization is meant an aggregation of matter which definitely is not in equilibrium with its surroundings. It exists in such a state by virtue of its own ability to preserve it. This distinguishes such entities from matter in a highly ordered form but with limited ability to preserve its structure, such as a crystal.

We are most concerned here with extra-terrestrial life which possesses intelligence. Again we have a difficulty in defining precisely what is required of a being for us to say it has such a faculty. We might require it to have the power of reason, or the ability to invent, create and imagine. More specifically, we would expect the existence of a set of aptitudes including that of foresight, imagination, learning and adaptability. These abilities have to be co-ordinated in such a way that the being can solve various problems posed for it in some fashion by its environment.

On earth humans have achieved the highest level of intelligence of all living species. Remarkably enough this has been obtained by apparently the maximum of economy in the type of basic building block being used. There is frugality at various levels, starting at the lowest in the most widely used elements – carbon, hydrogen, nitrogen, oxygen, and phosphorus, out of ninety-two possibilities in all – with carbon playing the major role. It then continues at molecular level with two basic types of large molecules constructed out of these six elements – proteins and nucleic acid. There are many forms of these molecules, but they are all basically re-arrangement of each other or enlarged versions of the same. Finally, these molecules are used in the fabrication of two types of cell, one of which is at a different electrical potential from its surroundings, and can transmit an electric current. It is the latter ability which has allowed intelligence to develop in animals, reaching its culmination in man.

The seat of human intellect is the brain, constructed with ten thousand billion excitable cells. The connections between this vast conglomeration are so complex as to still be beyond our understanding, and may be for some time to come. Yet some of the gross features of mental activity are becoming understood, and first steps have been taken in constructing machines which can act intelligently; one machine can even defeat the local draughts champion at his own game.

To develop to our level from the primitive chemical soup is thought to have taken several billion years. During that period it was constantly necessary to have precisely the right environment so that smaller, simpler molecules formed initially built up gradually to the more complex ones of proteins and nucleic acids, and these combined into single-celled animals, and so on. Given the same environment on another planet spinning round another star, it is to be expected that a similar type of evolution will occur. However, there are many steps in the chain leading to us, each apparently with many alternatives, and each depending very critically on the local conditions. It is likely that any intelligent life which does finally evolve on another reasonably similar planet may not be at all like us in physical form, even though it uses the same basic building blocks.

To appreciate the world through others' eyes we need first to ask what they will see of it. This will be critically determined by the environment in which they have evolved. If it is like ours we would expect a similar use of visible light, though nearby wave-lengths may be used, as in the case of the rattlesnake with its infra-red sensitive 'eye' enabling it to sense its warm-blooded prey in the dark. An environment considerably different from that on earth would naturally lead to a very different sensory equipment. But in order to have reasonable discrimination it would be necessary to observe with radiation of wavelength not too large. This could even be mainly by sound waves travelling through the planetary atmosphere, as is done on earth by porpoises in the water or bats in the air.

In our solar system the only planets appearing at first glance to be suitable for such an evolution of life are the Earth and Mars. Mercury has one side which is far too hot, the other far too cold and is devoid of atmosphere; Venus has far too hot a surface with a very dense atmosphere mainly composed of carbon dioxide with some nitrogen and water vapour; Jupiter is immensely cold and is made up mainly of methane and ammonia; Saturn is very similar, as are Uranus and Neptune, while Pluto is even colder, though with unknown composition.

It is possible that life based on ammonia, instead of water, may have evolved on the surface of Jupiter, which could be hotter than its outer atmosphere due to the latter trapping the sun's heat, the so-called greenhouse effect. Too little is known about the surface of Jupiter to be able to conclude on the possibility of life for sure; there may even be very suitable conditions. We will not rule it out here, especially as we will find later that intelligence developing on such a massive planet will have very different features from our own. This greenhouse effect might even allow life on Venus, though it would only be possible in the planet's cooler upper atmosphere.

Going further afield in our search for extra-terrestrial intelligent life, we have to look for a star which has lived long enough for life to have evolved to a high enough degree and one which would also have a planetary system. It is now thought that the process of stellar formation will in nearly all cases generate an orbiting system of planets. There is already direct evidence on this from nearby stars. Of the seventeen stars visible to us within twelve light years three of them are now known to have planetary companions. The most conspicuous example is Barnard's star (called so after its discoverer), which has a noticeable 'wobble' in its motion only explicable by the effects of an invisible companion about 50 per cent heavier than Jupiter.

The brighter component of the double star 61 Cygni also has an oscillatory motion, indicating it is accompanied by a planet of mass about eight times the weight of Jupiter. The other nearby star is number 21185 in Lalande's catalogue of heavenly bodies, it being just over eight light years distant, and appears to have a companion of about 1 per cent as heavy as the sun to explain its motion. This is getting close to the limit of planetary existence and near that for becoming a fully-fledged star; such massive planets may still allow life to develop.

The basic criterion for the evolution of intellect is thus that of stellar age. If we exclude multiple stars as creating too great a temperature variation for a common planet, it still leaves a reasonable proportion of interesting stars; of the one hundred and ten visible stars within twenty-two light years of

Earth it is estimated that there are fourteen good stellar candidates.

An attempt has already been made to receive communication from a civilization on one of these stars. It was conducted by Dr Frank Drake in the year 1959 in the United States, using an 85-foot radio telescope to detect non-random signals at the best wavelength. Two suitable stars were looked at, y Ceti and y Eridani, each about eleven light years away. Three months of watching gave no results, though we can only conclude that there was no signalling in our direction from these two stars during that period, and not that they were necessarily devoid of intelligence.

Rough calculations lead us to expect at least one million intelligent civilizations in our galaxy. Within a sphere of radius one hundred light years about the solar system there should be at least one or two such cultures. Our nearest neighbours need not be even as advanced as ourselves, yet among our galactic compatriots there must be many far more advanced than ourselves. What will they make of the universe surrounding them with problems?

The first difficulties they will meet will depend on their natural size. If it were possible for intelligence to arise in beings of only molecular proportions, their world would initially be totally different from our own. It would have an inherent probabilistic flavour, since they would continually be involved in quantum mechanical effects, none of which is certain. There would not be, for such beings, the logic we see in the world around us: either a thing is, or it is not. They would have a far wider range of possibilities: an event occurs with such and such a probability. Sense could only be made in their experience of many such events; single occurrences would not be relevant.

Assuming such minute beings could develop intelligence far superior to ours their further experiences of the world would involve larger segments of it than molecular-sized portions. They would realize the need for a certain Newtonian-type description of this gross world. To go further to the more distant problems of stars and then of galaxies would require enormous steps of reasoning and experiment, steps which would be almost impossible for them to make without evolution to a size suitable to realize the critical nature of gravitation and to withstand the natural forces around them. No doubt they could go far in understanding the complexities of the elementary particles, though even there they would ultimately be at a disadvantage, finding it very difficult to accelerate protons or electrons to very high speeds.

Our mini-beings would be very unlikely to discover the phenomenon of gravitational collapse; their sleep would not be disturbed by black holes appearing suddenly and swallowing them up. Such problems arise from effects acting over large distances. These would be more quickly noticed by beings more strongly affected by gravity. The heavier the planet the stronger is the gravitational attraction, so that intelligent life, if such can develop on a very heavy planet, will soon have realized the nature of the force ever pressing down on it.

There is no strong reason to doubt the possibility of evolution in an environment with several times the force of gravity, say, using the building blocks of earth-bound life – the proteins and nucleic acids mentioned earlier. They would naturally develop aggregations with greater strength than those on earth, so that single cells would have thicker membranes. There would be initially a high proportion of flattish-bodied life, but assuming that there was some dry land on the planet, animal structures would naturally develop great strength in order to move around on it. Gravity would be in their blood and soon in their understanding, so that intelligent life would soon have realized the nature of the all-pervading force on them. Indeed the nature of motion in high gravity would have been of great interest, so that would have been most explored.

The distortions of space and time by gravity would have been more natural to such beings than to ourselves. Not for them the extreme difficulty of puzzling over how to visualize a curved space or fathom time dilation. They would very rapidly have realized that gravity is geometry, and possibly gone far beyond even that.

The black hole would have been appreciated as a source of great energy and also of great danger. It could well have been created artificially, and even developed as a means of propulsion through space, giving the necessary high efficiency to blast free from their strongly-attracting planet. Such beings would be the masters of space and time, for they would have built into their very nature rapid responses to high accelerations and velocities and to curved motions. To them, moving on the earth would present little difficulty.

It is not clear that intelligent high-gravity beings would initially have much knowledge of the micro-world. Yet they would attempt to understand the properties of matter on their planet, and the sources of their sun's energy. This trail would inevitably lead them into the atom and even into its nucleus. With the greater power of gravity at their disposal their energy supplies would allow them to probe deep inside the elementary particles, so that the probabilistic world would rapidly be bared to their gaze. Superspace would then be discovered, so that the issue of the final state – of what happens to matter at the centre of a black hole – would become apparent.

The solution to the mystery of the black hole will be very important to the inhabitants of a planet close to the galactic centre. This is because various black holes will then be visible to them. One of these black holes will be the enormous one conjectured to be situated right at the galactic centre, being the source of Joe Weber's gravitational waves described in Chapter 5. This was expected to be one hundred million times as heavy as our sun, and so would be up to one light second across. An object of that size would not seem to be easy to observe, but its effects on stars even ten light years away will be as strong as that of the sun on Neptune; these stars will be in orbit around it. A 'solar' system more than ten light years across and, due to the higher stellar concentration near the galactic centre, involving millions of stars, would be very noticeable even up to a thousand light years away. The black hole would soon be an object of great interest, this even turning into fear if the inhabited planet were too close to the black hole for comfort. Life would not stretch ahead for the expected billions of years, till the death of their sun, but could be extinguished inexorably inside the black hole within a million years or less.

Once such a fate was realized as being in store the civilization would have to plan how to escape from their planet to find a more suitable habitation. A million years ahead would seem to be a long time; why worry about such a distant future? Yet all thoughts, all aspirations would be coloured by the knowledge that no one on the planet could live happily ever after. The quest for immortality would be hopeless, and a strong reaction against such a bleak future could arise when it was fully realized. Later, reason would reassert itself, and planning for escape would begin.

Civilizations near the galactic centre would have a second source of black holes, this being stars which had collapsed at the end of their nuclear evolution. Correspondingly, there would be more black holes than there are near us, again due to the higher concentration of stars near the galactic centre. This number would be even higher due to the greater age of the central stars than those in the spiral arms, with correspondingly greater chance of collapse. There could be as many as one thousand stars within ten light years of the life-giving planet's star, so as many, or more, black holes. The planet's inhabitants could well-nigh feel surrounded by black holes, pouring out X-rays and gravitational waves as their signature tune.

In either case the threat of being swallowed by a black hole would be sufficiently great to mobilize the civilization to escape. The problems to be solved to achieve this are fantastic. So far these difficulties have only seriously been considered by science fiction writers, usually following a fictional catastrophic supernova explosion of the parent sun. That danger is quite realistic, for a very massive planet would only be suitably close to a star for life to have developed if the star were itself considerably heavier than the sun, and so pull the planet close to it. Such a star could finish its life cycle in a period of hundreds of millions of years, to explode and then leave a supernova or a black hole. So if the heat did not get them, the black hole would.

The most difficult question for the threatened civilization to answer is that of energy; very large amounts would be needed to lift off an appreciable proportion of the members and contents of a civilization and give it any semblance of hope of finding a habitable planet elsewhere. The gravity-sensitive beings on a massive planet would have been able to find enough energy in gravitational collapse. Indeed it would be suitably ironic if they used the power of a black hole to escape the clutches of an even bigger one; no doubt they could escape from an exploding star in the same fashion.

As they travelled through space searching for a new home, the escapees would be able to call on their black-hole power unit at the most critical occasions, which is on close approach to a new star. Suitable scoops would be able to collect matter from interstellar space to feed into the voracious maw of their black-hole power generator. On approach to a star the available matter would increase enormously, both in the form of dust and gas as well as parts of planets. These increased sources of energy could help to power a complete planetary exploration as well as refurbish energy stores to prepare for journey to the next suitable star if no inhabitable planets were found around the present one.

The estimate we gave earlier, that about one-quarter of stars have planets, shows that such a voyage need not be a very long one. It would only have to continue till the threat of destruction had been overcome. That need only be of the order of tens of light years in the case of a supernova explosion of the parent star, and could be accomplished in that order of time. The journey would be at least ten times farther in the case of escape from the black hole at the galactic centre. But with an unlimited power supply in the black hole and such fuel as can be collected along the way such a trip, performed at close to the speed of light, would not prove insuperable.

High-gravity beings would very likely find our planet, if they ever visited it, only of passing interest. They would certainly be prepared to use it as a refuelling post, and would have noted the indigenous life form developing on it. They may have concluded that intelligence would not reach great heights due to the insufficient gravitational forces naturally occurring. They could even have preferred the more massive planets – Jupiter or Saturn – but may have found them too inhospitable, since too distant from the sun.

It is possible that the time-distortions of the black hole could be used by these alien beings to slow down the passage of time for themselves as they travel to the stars. We all of us desire immortality, paradoxically enough in a universe which may already be composed of 98 per cent dead matter. To modify a well-known phrase, 'in the midst of death we are alive'. It was already noted in Chapter 2 that this desire for immortality had no simple origin in early history. Various puzzling features were related to this question – the recorded longevity of the ancients, the so-human relationship between the 'sons' of God and mere men and women, the very experience of God itself, as well as various accounts of space-flight and similar experiences. Even more puzzling were the accounts of such happenings from different parts of the world: for example, Quetzalcoatl, the Feathered Serpent or Morning Star, was said to have instructed the Central American Indians in the sciences of agriculture, astronomy and architecture, and gave them a code of ethics. He was portrayed in the Codex Vindobonensis as descending from 'a hole in the sky' with the appearance of a space traveller.

A further puzzle arises over Satan, or the Devil. In the Old Testament, particularly the Book of Job, Satan roamed the earth looking for personal acts to report on adversely to God. The name Satan is, in fact, an English transliteration of the Hebrew for adversary. God was possessed of both good and evil; that which leads man into committing murderous acts or terrifies him comes from the lord. There was supposed to have been a mythical fight between God and the dragon of chaos, and only later did Satan play the role of the chief opponent to God. In the New Testament it was written 'God did not spare the angels when they sinned but cast them into hell' (II Peter, ii, 4). This type of revolt also occurs in other, even earlier, mythologies, such as in the Babylonian legend of the creation of the world, wherein the good god Marduk cleaved the goddess Tiamat in twain, and of the halves moulded the heaven and earth; after which he cast her henchman Kingu out of heaven and had him bound, and from his blood fashioned mankind, while Kingu's demons he confined to a dark place.

In the biblical legends why did God not destroy the Devil completely? And why was it necessary to throw the renegade angels into a certain place which was a 'fiery furnace'? One interesting explanation is that Satan and his followers were aliens disobedient to their leader and were punished by him by being cast into the black hole power source. As such they would appear to fade from view, but their image remain frozen against the event horizon of the black hole. The fiery appearance would come from the various radiations emitted in the process of their destruction.

Such a fate would have been final, with no reprieve. Yet stern justice would be necessary on such a voyage, and it would also act as a warning to mere humans who may have witnessed the scene. The horns usual in representations of the Devil could indicate antennae, confirming his similar origin to the visitors from the heavens of other cultures. So the concept of evil, as an idea, may have been brought from distant parts of the galaxy, along with other aspects of human culture.

One further interesting feature of past religions is that Saturn has always been attributed a leading role in determining man's fate. The god of time was Cronus – Saturn, in whose hands was the destiny of the world. Saturn was worshipped under the name of Ptah in Memphis in ancient Egypt; in China, Saturn was the Imperial Star. Yet it is not the brightest planet, and it is mysterious why it was given such a leading position amongst the Pantheons throughout the world. Why not the brighter Venus or Jupiter? Yet Saturn is second in size only to Jupiter in the planets of the solar system. It is a massive planet suitable for high-gravity beings. Is it that Saturn proved a more hospitable base from which the alien beings could explore the solar system?

If this conjecture is true then Saturn will have considerable surprises in store for a space-probe when it is sent there. At present we expect Saturn to have a similar structure to Jupiter: a dense atmosphere of hydrogen and helium, a shell of water-ice, a compressed inner shell of metallic hydrogen and a compressed rocky core. The greenhouse effect we mentioned earlier, together with being one-third the weight of Jupiter and twice as distant from the sun, may make Saturn's surface one just hospitable to life from denizens of nearly as massive a planet elsewhere in the galaxy. Saturn may be more suitable to them than the possibly hotter surface of Jupiter and certainly hot surface of Venus, at over nine hundred degrees.

Whether or not Saturn was a way-station of the odyssey of an extra-terrestrial civilization, it is still possible that our long-lived forefathers described in the Bible and discussed in Chapter 2 came from alien civilizations who had discovered how to tame black holes effectively. At the same time the origin of man's experience of God, and especially of the sons of God mentioned earlier, would be seen as arising from contact with such travellers from distant stars who had conquered the secrets of black-hole power generation. The resulting control of time by the aliens would have been interpreted by the natives as immortality. If there is any way of ultimately taming the black hole it is certainly to be expected that some of the supposed millions or so intelligent communities in the galaxy would have done so.

The lack of really hard evidence – pieces of space-craft or other artefacts – might lead one to think that such visits from outer space have not occurred in the past. Yet if only a handful or less have occurred over the last hundred thousand years, we do not expect to see a mound of rubbish left by the visitors on their voyages here. We can only expect the rare visit and search for it in the past records. We remarked earlier that there are descriptions of suitable occasions when such powerful visitors may have arrived. We can only wait and see if they come to us again before we have destroyed ourselves, either by our puny toys of atomic or chemical weapons, by our own garbage, or worst of all, by our experimental black holes. Let us hope that they arrive in time to teach us how to control the latter.

12 The Implications for Man

We have come to the end of our story about the universe. It is full of violent actions and grim forebodings, of horrors unfolded and mysteries still to be explored. And over all hangs the ultimate threat of being destroyed in either the neighbouring or universal black hole. It involves times which are vast compared to the human life span, and distances which dwarf man almost to nothing, making him a mere grain of sand on the beaches of the world.

The natural reaction to such a tale is that it is of sound and fury, but signifying nothing for man or woman. Each of us can continue to live our lives untouched by these immensities and by the catastrophes to come. The satisfaction gained from the simple round of life need be unaltered even when seen against this vast backdrop of the universe. We may live and die without raising up our eyes to the heavens, secure in the safety of our cotton-wool globe.

Yet that is false. We cannot divorce our lives from our understanding of the world around us, and especially of the basic problems of existence, the impossible questions of the universe. It is the answers, or lack of them, which determine our actions, even from day to day. For whatever we do, we must somehow come to terms with the infinite before we can act.

This can be seen by the fact that when any one of us performs some action, such as putting on his shoes or choosing a job, he usually is trying to achieve some sort of goal. It is evident what that goal is when he is getting his shoes on – it may be to go outside to achieve some further goal, say to play a game needing those special shoes, and so on. The higher-level goals are usually not so easy to justify – why does he take a job as a shoe-salesman rather than as an undertaker or a scientist?

Very often the highest-level goals are accepted unquestioningly. But yet they are there. They may have been derived from parents, or from school or friends, but ultimately they are based on the wish to survive and for loved ones to survive. This is the highest-level goal of all. Without it comes suicide, the end of all goals. Survival can be an end in itself or can be the means to a further end, especially for the religious. But the wish for survival, in one form or another, is absolutely essential for our continued existence.

This desire to persist in one's experience of the world takes many forms, the most important being that of immortality. To the Indian mind, be it Hindu, Buddhist, Jahnist or even one of the faith of Sikkhim, reincarnation is the most prevalent manifestation of the belief in immortality. This offers an explanation of the obvious inequalities into which men are born in this life, and a way of alleviating them by suitable rewards in the next one. For the life we live today arises from the law of karma, that the consequences of one's attitudes and conduct in a previous life determine the level of this one as inevitably as cause follows effect in the physical world.

To the Christian the idea of immortality is also essential: the soul is separate from the body and is indestructible. This idea goes back to early Greek philosophers, especially Plato, and has been accepted by many other religions as a basic tenet. It is by means of communion with God that man can have a wider vision of the world, and can even attain immortality if his conduct is satisfactory.

To the atheist there may still be basic processes which rule man's possible development in the world, even though such activities only involve physical entities. The most important of these rules governing the universe would appear, from the number of people accepting it, to be that of dialectical materialism, the doctrine that matter exists independently of thought and that it develops by successive oppositions or negations. This continual dialect or opposition has proved especially attractive when applied to the class struggle, leading to a situation in which one-third of the peoples of the world are required to profess belief in it.

How do these basic beliefs of survival, immortality and

dialectical materialism stand up to the story of the physical world described so far? We immediately see that survival as a short-term process, say over the next decade or so, need not be unduly affected by the advent of the black hole. Man and woman will still be able to go on their thoughtless daily round, unperturbed by such esoteric subjects as gravitational collapse, if they so wish, and no harm apparently be done. It would be dangerous to predict much beyond that time because of the speed of scientific developments today: looking back to thirty years ago – 1942 – and comparing it with today – atomic and hydrogen bombs, space ships sent to the moon and the planets, the dawning of the age of the computer, and many other changes – we see that scientific change progresses too rapidly even to begin to hint at where science and its applications will be in 2003.

Survival may well be threatened even before then by the weapons which science has placed in the hands of its political leaders. But we have learnt how to live with threats of nuclear and chemical warfare and still put out the cat and milk bottles and get a good night's sleep. Survival, as such, will only be severely threatened if experiments are performed to see if a black hole could be made. Such a test would have to require enormous advances in power generation and the understanding of gravitation. Yet when Einstein put forward his equation 'E=mc²' in 1905 he never realized that it would be used in 1945 to annihilate hundreds of thousands of his fellow men. So it is necessary for all of us seriously interested in survival to be constantly making sure that such possibilities have not occurred. There are very few, if any, mad scientists, but if the technical know-how to make a black hole became generally available how would a madman or fanatic be stopped from making it? If a small one made on earth was carelessly – or purposely – dropped it would sink rapidly to the earth's centre, where it would promptly proceed to devour the earth with great violence. There would be absolutely no way to stop it.

Here we come to the basic problem of future research in this area. The black hole is the ultimate doomsday weapon, as I mentioned before. How can we know when the corner is suddenly turned and the ability to make a black hole relatively easily is discovered? Should research in this area be put under security cover, as was done in the case of atomic research? It is no use the researchers in this field protesting that it would always be impossible to make a back-room black hole, or even a front-room one for that matter. It is just not possible to say at this time what can or cannot be done in such an area several decades hence.

The story of atomic research indicates that if there is a way of creating the black hole quite easily then it could, and would, most likely be found out by any one of several teams of research scientists working more or less independently. Secrecy may even intensify the attempts of others to solve the problems put 'under wraps', and so cause even more rapid solutions to be obtained. But yet some form of control must be prepared for just such an emergency as envisaged, of accidentally 'dropping' a man-made black hole. This watchfulness would at least save mankind from total annihilation, even if it would not necessarily save his complete freedom to do whatever he wanted. Even now research teams in the U.S. and U.S.S.R. are trying to condense matter a million-fold to white dwarf densities. Who knows when this will stop?

Such an extreme possibility as actually building black holes is far from the minds of scientists working in the field of gravitational collapse. To them it is an exciting and very disturbing area, where challenges are presented which cause their usual world to be swept away, to be replaced by topsy-turvy land. Yet they could ultimately cause our real world to be truly swept away and swallowed up by a black hole. It would then be very topsy-turvy indeed! All their research points to that end. It is up to mankind to be sure that the ultimate scientific experiment doesn't get carried out. And by that I mean each and every one of us, and not just the mythical 'common man' who is never oneself.

Immortality evidently suffers far worse than survival at the event horizon of the black hole. The fate of the physical universe is catastrophic, as we saw. It is either to be crushed into its fundamental constituents, as far as possible, to make a universal black hole, or it is to be slowly absorbed by local black holes, again to be crushed out of existence as we know it. Long before that has happened our physical bodies would have become unrecognizable, so that at such an end we can only appeal to our souls to preserve us. They would have to keep well clear of the black holes of the universe if they are to continue any form of existence, though of course this is an unnecessary warning, since souls are not supposed to have any physical position in the first place.

It could only be if the universe bounces back again after its collapse that these separated souls have any chance of returning to matter which could realistically be said to be worth having a soul. There is very little evidence of such a bounce being able to occur, but if it does, only then can one expect any form of immortality. That this is the only possibility for everlasting existence depends further on the condition that the re-expanding matter so reconstructs itself that an effective mind-body interaction will be possible once again. Such a future development of matter need not occur after the bounce, but at present we have no knowledge of how to discuss this further.

The only chance of immortality then is in an oscillating universe. Even in that, everlasting life will not be of the usual form but one in which there may be no relation at all between one cycle and the next due to the enormous re-scrambling of matter in the collapsed phase. It could well be that souls will have to cast lots as to which of the variety of bodies they will inhabit on subsequent lives. That is, of course, unless the hand of God intervenes, his wonders to perform. But in that case there is nothing to discuss, only to believe.

The other possible fates of the universe involve a true death of all matter through final disappearance inside one or several event horizons, never to reappear. It can be conjectured even further, however, that immortality can be attained by falling through a rotating black hole into its next universe. That might be regarded as almost on the same level as the immortality gained in the oscillating universe, yet it is still to be regarded as true death in this. One's closest friends may well be captured by a completely different black hole, and so be driven into another universe altogether. So there is always the chance that the immortality gained by falling through a rotating black hole may be a very lonely one.

Even if there is no such thing as a soul there is still a faint chance of immortality in an oscillating universe, or of further existence in a different universe after falling through a rotating black hole. For re-expanding matter, in the former case, might still develop into a state in which living forms develop and even have associated mental capacities. There could even be another you reading this page in the next universe, the page, the book it is in and the John Taylor who wrote it having all been created in the next universe. But the chance of this occurring would seem to be infinitesimal; it would require the millions of processes which occurred in just the order and speed they did to produce you, me and the book here and now to be repeated almost identically in the next oscillation of the universe. Even the recreation of our Milky Way exactly in its present form would be unlikely in the next universe, let alone the reconstruction of our solar system, yet there would probably be habitable planets with intelligent life forms on them. Life will very likely still go on in such an oscillating universe, but will not know anything about its detailed forms in preceding or subsequent versions of the oscillations.

Survival, immortality, and now dialectical materialism. In the true black hole, with no bounce or re-expansion into another world, we see there is only a one-way process, always to collapse. It is only if there can be re-expansion, either through the spin of the black hole or by means of the unknown forces in the bouncing universe, that any opposition to the force of gravity can be recognized. But such a two-way process occurs only over cosmic times, and the efforts of men to see this mirrored in their own lives will always be doomed to failure. Societal interactions are too complicated to be reduced to the grand simplicity of the universe; it is like expecting the activity in an ants' nest to be as simple as that of a frictionless ball rolling on a perfectly smooth plane.

Black holes have important implications for mankind on these levels of survival, immortality and dialectic. They have further relevance to man's quest for meaning when faced with the infinite. The answers provided by the black-hole universe are simple in their outlines, but they are not yet final. As has been emphasized before, there may come a time when such answers are regarded as being the last word, though even that would never be certain.

Whether or not such finality appears to be being reached, the advances made in our understanding of the mysterious universe by the black hole are undeniable. It effects such extreme distortions of space and time that their basic properties are thrown into clear relief. But the black hole itself brings with it further difficulties which, in fact, also illuminate our concepts of time, space, causality and the beginning and ending of the universe. In terms of these clarifications it is even possible to understand various aspects of man's developments in the past. The power given by the black hole to alien civilizations could explain many curious features of early records – the Gods, and men's relations with them, the development of early civilizations, experiences of immortality, the nature of Satan and evil, the exalted position of Saturn in the Pantheons of early mythologies, and so on and so on.

Beyond these specific ways that the black hole universe appears relevant to everyday actions and the historical past is the broader one of giving a completely new view of life from many aspects. There are numerous interesting points here on which one can dwell, for example, the manner in which time travel into the past would be possible inside a rotating black hole or anywhere in space if a naked singularity could exist, or the light which gravitational collapse throws on the beginning and end of the universe, and puzzles about the edge of the universe, and so on.

One of the most surprising of these is that our understanding of the world in the very small indicates that it has a probabilistic character. In order that such a chancy world makes sense it is necessary that one of two possibilities must be true. The first is that there must be a plethora of universes, all developing in time according to the laws we are beginning to understand, but all different in some way from each other. The other is that the universe must be infinite in extent.

The first case has some very bizarre consequences. For example in one of these many universes you might still be reading this page but it will have a blue colour to it, while the other pages of the book are white. Another universe might have all the other pages of the book blue, but this one white, or in a different world it might still be an identical book but by a completely different author.

It is quite remarkable that from the micro-world we can predict the possible existence of this host of other macro-worlds. We will never be able to interact with them, at least according to the present theoretical understanding of this situation. I would never be able to meet my identical twin in the countless other worlds in which he, or rather I, exist. I will be experiencing different phenomena in these other worlds, so in some sense will be different from the I in this world. There will also be universes in which the person I have become will be so different as to be difficult to recognize with certainty.

Indeed it is possibly a good thing, for our own peace of mind, that we cannot explore these other worlds directly. Meeting my identical twin in all things but his ability to fly or walk on the ceiling could give me quite a shock. It is better for all of us that these bizarre worlds in which 'we' also exist are denied us. Similarly no replicas from other worlds can come through to us, certainly to be disturbed in their turn and possibly to give us a scare as well.

Yet here again the black hole may intervene as the gateway to one or other of our related worlds. Indeed if matter ever again comes out of collapse in a black hole into another universe it may sometimes do so in a manner which would tend to develop into a similar configuration from which it collapsed. This is a reasonable result if we assume that the collapse and expansion do not go through too highly a condensed intermediate state, so that not too much information about the previous world is destroyed. This could be achieved by a large rotating black hole, say greater than a million times as heavy as the sun.

In this fashion we may kill two birds with one stone, and identify the set of probabilistic worlds our understanding of the universe requires with those joined on to a spinning black hole. Such an equality may be incorrect, but has the advantage of reducing the multiplicity of worlds needed. We will not put much faith in it here.

Whether or not we can achieve such an identification, each of our black-hole universes is one among many. We have finally reduced man to the most lowly position possible. Not only does he live on an average planet of an average star in an average galaxy, but his whole universe is only expected to be average among the multitude of others. That is the final let-down of all. Nor can we put all these universes together to make a big one, as we tried to do just now. For the resulting collection of universes is no longer a universe of the same sort at all. These worlds are really to be regarded as alternatives to each other, not as additions. So we are really and truly stuck with our average position among all these alternatives.

The second case, that there is a universe of infinite extent, means that there must always be mystery, since we can never hope to investigate it completely. But certainly we do not expect to be in any special position in this infinite universe.

Either possibility has very strong bearings on the way we should see ourselves. The collection of the mysteries of the universe is conventionally called God. We, as intelligent beings, appear to occupy absolutely no privileged position with regard to God. It (I don't wish to denigrate the female of the species by calling God 'him', nor can I call him 'she' because God must in any case be asexual) can have no 'interest' whatsoever in our developments. For God has no human attributes of any kind. How can the black hole of the universe, its microstructure, and its infinite reaches relate, in any human fashion, to a group of very small aggregations of energy, each of which is called a human being?

The constant rejoinder to such a position is that we are singled out by our minds or 'souls'. Let us accept that such independent entities do, in fact, exist. Lower animals, especially the chimpanzees, can be regarded as being as well qualified as we of having minds, at least from recent results on their language ability. So mankind is not singled out by the possession of a mind. In any case we expect the existence of similar intelligent beings on other planets round other stars, even in other galaxies. In this gathering of intelligent beings is it 'mind' alone which is to be singled out as having any possible relation to God? In the black-hole universe, that is the best that can occur; even that may not be enough, and then we can only be complicated machines.

We recognized earlier that the persistence of mind is only of interest in a bouncing universe, either coming out from a non-rotating collapse into the original universe or expanding into another universe in the spinning case. In the other possibilities of ultimate expansion or of one final, non-rotating collapse, mind will in any case finally have no active matter to hook up to.

In order for mind to continue its existence through the bounce its connection to matter must be either completely destroyed during the collapse or be tenuous enough as to retain its information without collapse. To achieve this, minds cannot possess any positive energy, since otherwise they would be crushed by the collapse, along with all other forms of energy. Mind must thus be completely detached from matter during the fateful bounce.

There is one quantity which can escape a black hole, and that is a tachyon (from the Greek tachys, meaning swift) since it is, by definition, a particle which can travel faster than light. To do so it must have imaginary mass, and certainly can never be slowed to a speed equal to that of light. This would seem to make a theory of mind based on tachyons very difficult indeed. We are left, then, with a complete decoupling of mind during collapse. Of course if mind has an independent existence then in any case it has to decouple from matter at death pretty strongly, so that such a process need not be impossible on this account.

The complete independence between mind and matter in the bouncing black-hole universe makes it very difficult to understand how such a coupling can be achieved in any realistic manner: there is absolutely no physiological evidence of a region in the brain acting as a centre for such coupling. It is one of the mysteries we put into God if we have such a strong need for an immortal soul. Otherwise we have to assume that mind dies along with structured matter in the bounce. It will re-appear in the expansion phase, but only, in all likelihood, as an aspect of matter.

The black-hole universe certainly affects the way we can understand life and ourselves. It holds out little hope for immortality, but great opportunities to expand our vision of the world during this life. We can never say, even now we see some of the outlines of the fate in store for us, 'Stop the universe, I want to get off', though some of the brave among us might say 'Stop the universe, I want to dive through'. But we can only hope that as the future looming mightily ahead of us becomes more clearly defined we can recognize its face more clearly. It is that of ourselves and all those around us – we are at one with the universe since we are energy manifestations, as is the rest. Our oneness with the world will allow us to accept whatever fate it will have in store for us. We need not be frightened by other aspects of ourselves. We must be humble in the face of the laws of the world, and especially those governing the black-hole universe. Our ultimate fate is collapse. Beyond that is the still unknowable part of infinity.

Postscript

In the year since this book was written, there have been a number of important developments in black-hole physics. They affect the evidence for the existence of black holes of various shapes and sizes, the sorts of properties that black holes may have, and also the difficulties that black holes present. There is an even deeper result which deals with the very possibility of the existence of black holes themselves.

Experimental evidence was presented in Chapter Five for the existence of black holes. One possible candidate was the variable X-ray star, Cygnus X-1. Further results have become available on the X-ray emission from this star, but we have not been able to observe with enough detail the variations which would prove that Cygnus X-1 must be a black hole. On present evidence it could still be a neutron star. There will undoubtedly be more refined observations which will allow the difference between these two possibilities to be finally settled.

On the theoretical side, there have been suggestions which would allow the irregular variations of the X-rays from Cygnus X-1 to be explained without invoking a black hole. One of these was a suggestion that the invisible secondary of Cygnus X-1, which was thought to be the source of the X-ray variations, is a neutron star with a very strong magnetic field around it. Due to the rotation of this field, varying amounts of radiation can be emitted as matter falls into the neutron star.

An alternative suggestion has been that Cygnus X-1 is not simply a binary star-system, but is actually a triple star-system. The condensed invisible object which gives rise to the rapidly varying X-rays is supposed to be rotating round a more diffuse secondary which itself rotates round the visible primary. This suggestion makes very clear predictions as to the nature of the X-ray variation and will, no doubt, be able to be tested and proved within the next year or so.

In any case, there are at least five other variable X-ray stars which have recently been found and which would appear to be good candidates for black holes. Similar careful investigations of the nature of their radiation will be necessary before they will be able to be finally accepted as bona fide black holes or not.

One of the other important pieces of evidence for the existence of black holes was also discussed in Chapter Five – the gravitational radiation which Professor Weber had been detecting by his large aluminium cylinders acting as gravitational wave-detectors. Similar apparatus has been set up by numbers of other groups since that time, and in the past year they have been obtaining results. Not a single group has been able to duplicate Weber's observations, and indeed only one or two possible candidates for pulses of gravitational radiation have been observed during running periods of two or three months. It may be, therefore, that Weber has not been observing gravitational radiation itself, but some effects that may involve electromagnetic radiation, for example, arising from the atmosphere. This explanation of the disparity between Weber and the other experimental groups is not absolutely necessary at this stage, since the various other groups do not have equipment identical to Weber's.

Thus, the situation would appear to be a little unclear at present, and it may still be that what Weber is observing is indeed gravitational radiation. Undoubtedly the disagreements will be resolved within the next few years when much more sensitive equipment – at present under design – will be in operation. The sensitivity will be increased by a factor of at least a hundred in the new equipment, and should even permit detection of gravitational pulses from supernovae explosions in the galaxy.

There is one difficulty about this, which is that there is only about one suitable explosion every thirty years in our own galaxy. To obtain an acceptable event rate it will be necessary to detect supernovae explosions in nearby clusters of galaxies. An example of this is the Virgo Cluster, which contains over a thousand galaxies, and might produce several supernovae explosions per year. Though this is an acceptable rate of production of pulses, it would still require an increased sensitivity of 100 million over present-day apparatus. Even so, it might be possible to achieve this by using suitable extensions of methods at present under development.

In Chapter Six, various properties of black holes were discussed. A particularly interesting possibility has recently been developed by Steven Hawking, in which he has suggested that black holes may even annihilate themselves by creating a very large number of negative energy particles near the surface of their horizon, which then fall in and destroy the black hale itself. The rate of production for large black holes, as massive as the sun or more, is such that their lifetime for such self-destruction is much longer than that of the universe. However, this is not true for many black holes, say, of the size of a ten-thousandth or a hundred-thousandth of a gram. Such mini-holes would destroy themselves within a second or so in a very violent explosion, at least according to Hawking's estimates.

Such a possibility is of great interest, though more recent calculations indicate that only the spinning energy of such mini-holes will be radiated away and they will be left without spin. This is an area of great interest and no doubt, much more work will be done on this and the picture clarified within the next few years.

The book contains considerable discussion of the difficulties which the existence of black holes presents to both science and to mankind in general. One of the scientific problems raised in Chapter Five was the loss of neutrons and protons in a black hole; the nucleon is not supposed to decay in any way, and indeed has not been observed to do so in laboratory experiments, even though very careful tests have been made. To date, the lifetime of a proton is thought to be at least 10^{21} years. However, very recently it has been suggested that protons and neutrons can indeed decay to electrons. This decay can be explained in terms of the new theories of unified weak and electromagnetic interactions, and leads to a decay lifetime for the proton of about 10^{28} years. This is in agreement with present experimental data, though the experiments are being redesigned to test in detail the predictions of this model. If nucleons do decay, then the threat presented by the black holes to the nucleons existing outside it need not be regarded as disastrous as all that; they will decay by more conventional methods in any case.

The most crucial difficulty presented by black holes is the problem as to what happens to matter at their centre. This question has still not been resolved, nor may it be for a long time to come. It has certainly now been fully realised as one of the most crucial problems of science today, and a recent conference on quantum gravity at Oxford (February 15-16, 1974) made this abundantly clear. As Professor John Wheeler, possibly the 'father' of the black hole, remarked at the Conference, it may be necessary to give up the theory of gravitation in order to be able to deal with this particular problem. A number of contributions from other scientists indicated that it is not possible to marry quantum mechanics with the theory of gravitation if the latter is that of Einstein. Such a result is not disturbing to those who wish to unify the four forces of nature – those of electricity and magnetism, of the nucleus, of radioactivity, and of gravity. But yet, to design a new theory of gravity which will fit together better with the uncertain world of matter is a very big job. We certainly do not know how to do that yet.

Without doubt, black holes will still be able to be produced in any new unified theory of the forces of nature. This is clear for large black holes, say of the size of a hundred million solar masses, since whatever new effects arise in highly compressed matter will not affect its behaviour at the rather low densities occurring as such an object collapses inside its event horizon.

If and when such a theory is produced, it may well be that though black holes are still required, what happens at their centre will be very different from what we are at present conjecturing. It will still leave us in the difficult position of never being able to verify our predictions. This raises a question mark against the efficacy of the scientific method – one which we may never be able to erase.

Bibliography

SIR FRED HOYLE. *Of Men and Galaxies*, Heinemann, London (1964)

G. ABELL. *Exploration of the Universe*, Holt, Rinehart & Winston (1964)

SIR JAMES JEANS. *The Stars in their Courses*, C.U.P. (1954)

GEORGE GAMOW. *The Creation of the Universe*, Viking Press, N.Y. (1961)

T. PAGE and L. W. PAGE. *The Evolution of the Stars*, Macmillan (1968)

T. O. PAYNE. 'Man's Future in Space', *Contemporary Physics* (Taylor & Francis Ltd.), 13, 393-401 (1972)

YA. B. ZELDOVITCH. 'Survey of Modern Cosmology', *Advances in Astronomy and Astrophysics*, 3, 242-379 (1965) (Academic Press)

J. N. BAHCALL. 'The Solar Neutrino Problem', *Comments on Nuclear and Particle Physics* (Gordon & Breach), 5, 59-64 (1972)

W. A. FOWLER. 'What Cooks with Solar Neutrinos?', *Nature*, 238, 24-26 (1972)

P. W. HODGE. *The Physical Astronomy of Galaxies and Cosmology*, McGraw Hill, N.Y. (1966)

W. BONNOR. *The Mystery of the Expanding Universe*, Eyre & Spottiswoode, London (1964)

D. W. SCIAMA. *Modern Cosmology*, C.U.P. (1971)

The Structure and Evolution of Galaxies, Interscience Pub. N.Y. (1965)

H-Y CHIN, R. L. WARAIHA and J. L. REMIO. *Stellar Astrophysics*, Vols 1 and 2, Gordon & Breach, London (1969)

J. LEQUEUX. *Structure and Evolution of the Galaxies*, Gordon & Breach, London (1969)

G. and M. BURBRIDGE. *Quasi-Stellar Objects*, W. H. Freeman & Co., San Francisco (1967)

A. G. W. CAMERON. *Interstellar Communication*, W. A. Benjamin Inc. (1963)

S. H. DOLE and I. ASIMOV. *Planets for Man*, Methuen Co. Ltd. (1965)

C. SAGAN and S. SHKLOVSKII. *Intelligent Life in the Universe*, Holden Day (1971)

G. J. WHITROW. *What is Time?*, Thames and Hudson (1972)

P. G. BERGMANN. *The Riddle of Gravitation*, John Murray (1968)

E. L. SCHATZMANN. *The Structure of the Universe*, Weidenfeld & Nicholson (1968)

R. RUFFINI and J. WHEELER. 'Introducing the Black Hole', *Physics Today*, June (1971)

W. ISRAEL. 'Gravitational Collapse and Causality', *Physics Review*, 153, 1388 (1967)

F. DE PEAT. 'Black Holes and Temporal Ordering', *Nature*, 239, 387 (1972)

V. DE LA CRUZ, J. E. CHASE and W. ISRAEL, 'Gravitational Collapse with Asymmetries', *Physics Review*, Letters, 24, 423 (1970)

K. S. THORNE. 'Gravitational Collapse', *Scientific American*

J. G. TAYLOR. *The New Physics*, Basic Books, N.Y. (1972)

J. G. TAYLOR. 'Requiem for the Universe', *The Listener*, 86, 2215 (1971)

G. WICK. 'The Clock Paradox Resolved', *New Scientist*, 3 Feb. 1973, p. 261

G. J. C. HAFELE and R. E. KEATING. 'Around the World Atomic Clock', *Science*, 177, 166 (1972)

F. J. DYSON. 'Energy in the Universe', *Scientific American*

J. G. TAYLOR. 'Particles Faster than Light', *Science Journal*, Sept. 1969, p. 43

R. PENROSE. 'Black Holes', Chapter in *Cosmology Now*, B.B.C. Publications, 1973

D. LYNDEN-BELL. 'Cosmic Power', Chapter in *Cosmology Now, B.B.C. Publications* (1973)

D. Sciama. 'Models of the Cosmos', Chapter in *Cosmology Now*, B.B.C. Publications (1973)

W. H. McCrea. 'The Problems of the Galaxies', Chapter in *Cosmology Now*, B.B.C. Publications (1973)

M. Rees. 'The Far Future', Chapter in *Cosmology Now*, B.B.C. Publications (1973)

J. G. Taylor. 'The Implication for Man,' Chapter in *Cosmology Now*, B.B.C. Publications (1973)

A. Tomas. *We Are Not the First*, Souvenir Press Ltd. (1971)

J. G. Taylor. *The Shape of Minds to Come*, Michael Joseph (1971)

F. Hoyle. *Frontiers of Astronomy*, Harper & Bros. (1955)

W. C. Hernandez, Sr. 'Kerr Metric, Rotating Sources and Machien Effects', *Physics Review*, 167, 1180 (1968)

B. B. Godfrey. 'Mach's Principle, the Kerr Metric and Black-Hole Physics', *Physics Review D* 1, 2721 (1970)

R. Penrose. 'Gravitational Collapse and Space-Time Singularities', *Physics Review*, Letters, 14, 57 (1965)

D. Christodolon. 'Reversible and Irreversible Transformations in Black-Hole Physics', *Physics Review*, Letters, 25, 1596 (1970)

W. Press and S. A. Teukolsky. 'Floating Orbits, Superradiant Scattering and the Black-Hole Bomb', *Nature*, 238, 211 (1972)

R. K. Pathra. 'The Universe as a Black Hole', *Nature*, 240, 298 (1972)

P. Vodzis, H-J Seifot and H. Muller zum Hagen. 'On the Occurrence of Naked Singularities in General Relativity', Hamburg Univ. Report, March 1973 (unpublished)

G. W. Gibbons. 'On Lowering a Rope into a Black Hole', *Nature*, 240, 77 (1972)

J. D. Beckenstein. 'Baryon Number, Entropy and Black-Hole Physics', Princeton Univ. thesis (unpublished, May 1972)

S. W. Hawking. 'Gravitationally Collapsed Objects of Very Low Mass', Mon. Not. R. Astr. Soc. 152, 75 (1971)

C. T. Bolton. 'Dimensions of the Binary System HDE 226868=Cygnus X-1', *Nature*, 240, 124 (1972)

G. W. Gibbons and S. W. Hawking. 'Evidence for Black Holes in Binary Star Systems', *Nature*, 232, 465 (1971)

J. R. Gott *III*. 'Further Evidence for Collapsed Objects in Binary Star Systems', *Nature*, 234, 342 (1971)

E. N. Walker. 'Band V Photometry of Cygnus X-1', Mon. Not. R. Astr. Soc., 160, 9P (1972)

B. L. Webster and P. Murdin. 'Cygnus X-1 – a Spectroscopic Binary with a Heavy Companion?', *Nature*, 235, 37 (1972)

Index of Subjects

Index of Names

Some Modern Classics in Fontana

The Renaissance (Illus.)   Water Pater

Renaissance and Baroque (Illus.)   Heinrich Wölfflin

The Ancien Régime and the French Revolution   Alexis de Tocqueville

Varieties of Religious Experience   William James

The Problem of Pain   C. S. Lewis

Apologia pro Vita Sua   John Henry Newman

The Phenomenon of Man   Teilhard de Chardin

No Rusty Swords   Dietrich Bonhoeffer

Myths, Dreams and Mysteries   Mircea Eliade

The Courage to Be   Paul Tillich

Utilitarianism   John Stuart Mill

The English Constitution   Walter Bagehot

Man Makes Himself   V. Gordon Childe

Course in General Linguistics   Ferdinand de Saussure

Axel's Castle   Edmund Wilson

To the Finland Station   Edmund Wilson

A Fontana Selection

Ideology in Social Science, edited by Robin Blackburn
Stonehenge Decoded, Gerald S. Hawkins
Romantic Image, Frank Kermode
Memories, Dreams, Reflections, C. G. Jung
How to Write Reports, John Mitchell
Reformation Europe (1517-1559), G. R. Elton
Social Problems in Modern Britain,
   edited by Eric Butterworth and David Weir
The Screwtape Letters, C. S. Lewis
My Early Life, Winston Churchill
Voyage to Atlantis, James W. Mavor
Chance and Necessity, Jacques Monod
Natural History of Man in Britain, H. J. Fleure and M. Davies
The Book of Ireland Frank O'Connor (ed)
The Wandering Scholars, Helen Waddell
Italian Painters of the Renaissance, Bernhard Berenson